Assessment and Renewal of Water Distribution Systems

Awwa
Research
Foundation

About the Awwa Research Foundation

The Awwa Research Foundation (AwwaRF) is a member-supported, international, nonprofit organization that sponsors research to enable water utilities, public health agencies, and other professionals to provide safe and affordable drinking water to consumers.

The Foundation's mission is to advance the science of water to improve the quality of life. To achieve this mission, the Foundation sponsors studies on all aspects of drinking water, including supply and resources, treatment, monitoring and analysis, distribution, management, and health effects. Funding for research is provided primarily by subscription payments from approximately 1,000 utilities, consulting firms, and manufacturers in North America and abroad. Additional funding comes from collaborative partnerships with other national and international organizations, allowing for resources to be leveraged, expertise to be shared, and broad-based knowledge to be developed and disseminated. Government funding serves as a third source of research dollars.

From its headquarters in Denver, Colorado, the Foundation's staff directs and supports the efforts of more than 800 volunteers who serve on the board of trustees and various committees. These volunteers represent many facets of the water industry, and contribute their expertise to select and monitor research studies that benefit the entire drinking water community.

The results of research are disseminated through a number of channels, including reports, the Web site, conferences, and periodicals.

For subscribers, the Foundation serves as a cooperative program in which water suppliers unite to pool their resources. By applying Foundation research findings, these water suppliers can save substantial costs and stay on the leading edge of drinking water science and technology. Since its inception, AwwaRF has supplied the water community with more than $300 million in applied research.

More information about the Foundation and how to become a subscriber is available on the Web at **www.awwarf.org**.

Assessment and Renewal of Water Distribution Systems

Prepared by:
Neil S. Grigg
Colorado State University, Fort Collins, CO 80523-1372

Sponsored by:
Awwa Research Foundation
6666 West Quincy Avenue, Denver, CO 80235-3098

Published by:

Awwa Research Foundation
Advancing the Science of Water®

IWA Publishing

DISCLAIMER

This study was funded by the Awwa Research Foundation (AwwaRF). AwwaRF assumes no responsibility for the content of the research study reported in this publication or for the opinions or statements of fact expressed in the report. The mention of trade names for commercial products does not represent or imply the approval or endorsement of AwwaRF. This report is presented solely for informational purposes.

Published by IWA Publishing, Alliance House, 12 Caxton Street, London SW1H 0QS, UK

Telephone: +44 (0) 20 7654 5500; Fax: +44 (0) 20 7654 5555; Email: publications@iwap.co.uk;
Web: **www.iwapublishing.com**

AwwaRF report number 91025F. Originally published by AwwaRF for its subscribers in 2004.
IWA Publishing version published 2005.

ISBN 1 84339 914 8

CONTENTS

TABLES

FIGURES

FOREWORD

The Awwa Research Foundation (AwwaRF) is a nonprofit corporation that is dedicated to the implementation of a research effort to help utilities respond to regulatory requirements and traditional high-priority concerns of the industry. The research agenda is developed through a process of consultation with subscribers and drinking water professionals. Under the umbrella of a Strategic Research Plan, the Research Advisory Council prioritizes the suggested projects based upon current and future needs, applicability, and past work; the recommendations are forwarded to the Board of Trustees for final selection. The foundation also sponsors research projects through an unsolicited proposal process; the Collaborative Research, Research Applications, and Tailored Collaboration programs; and various joint research efforts with organizations such as the U.S. Environmental Protection Agency, the U.S. Bureau of Reclamation, and the Association of California Water Agencies.

This publication is a result of one of these sponsored studies, and it is hoped that its findings will be applied in communities throughout the world. The following report serves not only as a means of communicating the results of the water industry's centralized research program but also as a tool to enlist the further support of the nonmember utilities and individuals.

Projects are managed closely from their inception to the final report by the foundation's staff and large cadre of volunteers who willingly contribute their time and expertise. The foundation serves a planning and management function and awards contracts to other institutions such as water utilities, universities, and engineering firms. The funding for this research effort comes primarily from the Subscription Program, through which water utilities subscribe to the research program and make an annual payment proportionate to the volume of water they deliver and consultants and manufacturers subscribe based on their annual billings. The program offers a cost-effective and fair method for funding research in the public interest.

A broad spectrum of water supply issues is addressed by the foundation's research agenda: resources, treatment and operations, distribution and storage, water quality and analysis, toxicology, economics, and management. The ultimate purpose of the coordinated effort is to assist water suppliers to provide the highest possible quality of water economically and reliably. The true benefits are realized when the results are implemented at the utility level. The foundation's trustees are pleased to offer this publication as a contribution toward that end.

The goal of this project is to help utilities assess the knowledge they need for distribution system capital management and to evaluate the available research-based tools. The work focused on the tasks involved in renewing water infrastructure—locating, assessing, planning, repairing, and renewing water distribution systems. This report contains a synthesis of knowledge and the suggestions of utility and other professionals about renewing distribution systems. In addition to utilities, the report is aimed at policy makers and at the water supply research and development community (including AwwaRF), to help these professionals plan a responsive research agenda for distribution systems. It is hoped that the report will serve as a guide and a textbook for utilities seeking to minimize their exposure to risk and threats from all sources.

Walter J. Bishop
Chair, Board of Trustees
Awwa Research Foundation

James F. Manwaring, P.E.
Executive Director
Awwa Research Foundation

ACKNOWLEDGMENTS

The project's participating utilities and their representatives deserve the most credit for informing the investigators about their practices and in guiding the research. The authors of this report are indebted to the following water utilities and individuals for their cooperation and participation in the project:

- Kanwal Oberoi, Charleston Commissioners of Public Works, Charleston, S.C.
- Alex Margevicius, Cleveland Water Division, Cleveland, Ohio
- Billy Cobb, Columbus Water Works, Columbus, Ga.
- John H. Bambei Jr., Denver Water, Denver, Colo.
- Jim Hibbard, Fort Collins Water Utility, Fort Collins, Colo.
- Bret Russell, Louisville Water Company, Louisville, Ky.
- David Meza, Marina Coast Water District, Marina, Calif.
- Gary Shipley and Mark Bushyeager, Metropolitan Water District of Southern California, Los Angeles, Calif.
- Joseph Becker, New Orleans Board of Water and Sewerage, New Orleans, La.
- Darell Zimbelman, Northern Colorado Water Conservancy District, Berthoud, Colo.
- Dave Bries and John Tytla, Olathe Water Department, Olathe, Kan.
- Kelly Martin, City of Ottawa, Ont., Canada
- Charles H. Call Jr. and Mike Lewis, Salt Lake City Corporation, Salt Lake City, Utah
- John Galleher, San Diego County Water Authority, San Diego, Calif.
- Rodney Liesinger and Tony Everson, Sioux Falls Water Department, Sioux Falls, S.D.
- Fred Remen, South Florida Water Management District, West Palm Beach, Fla.
- Rudi Radovanovic, Sydney Water, Sydney, Australia
- Christine Owen, Tampa Bay Water, Tampa, Fla.
- Michael Woodcock, Washington Suburban Sanitary Commission (WSSC), Laurel, Md.

Others who made special contributions at the project workshop were

- Tom Vidmar, Homestake Project, Aurora, Colo.
- Tom Horn, Columbus Water Works, Columbus, Ga.
- Ray Olson and Jeremy Ross, Denver Water, Denver, Colo. (two workshops)
- Kofi Boateng, Washington Department of Water Resources, Washington, D.C. (two workshops)
- Michael D. Royer, U.S. Environmental Protection Agency, Washington, D.C. (two workshops)
- Patrick Clinton, East Bay Municipal Utility District, Oakland, Calif.
- Makoto Matsushita, Kobe Municipal Waterworks Bureau, Kobe, Japan (unable to attend workshop but contributed materials)
- George Mavrakis, Los Angeles Department of Water and Power (LADWP), Los Angeles, Calif. (also hosted focus group at LADWP)
- Leif Sigurd Hafskjold, SINTEF, Norway
- Gary Stine and Michael Stift, San Diego Water Authority, San Diego, Calif. (also met with investigator at San Diego Water Authority)

- Sharon Trost, South Florida Water Management District, West Palm Beach, Fla.
- Terry Van Dyke, Sioux Falls City Engineering, Sioux Falls, S.D. (also contributed photos for report)
- John Troutt, Tampa Bay Water, Tampa, Fla.
- Arun Deb, formerly of Weston Solutions, West Chester, Pa.
- Ahmad Habibian, Black and Veatch, Baltimore, Md.
- Paul Fanner, Water Systems Optimization, Miami, Ohio
- Yakir J. Hasit, Parsons, Philadelphia, Pa. (now at CH2M HILL, Philadelphia)

The project director would like to express particular thanks to Michael Woodcock of WSSC, who contributed substantially to the project from its inception to its finish, including presenting a valuable briefing to the project director and hosting a workshop at WSSC.

The authors also thank Peter Fitzwilliams, Dallas Water Utilities (Texas), and Diane Sacher, Winnipeg Water & Waste Department (Man., Canada), for attending the workshop and serving on the Awwa Research Foundation (AwwaRF) Project Advisory Committee. They also thank Jian Zhang, AwwaRF's project manager, and Kate Martin, the original project manager.

The project director also thanks the many persons who contributed information to the report during the information collection phase. They are too numerous to list, but they include representatives of government agencies, industry associations, private firms, utilities, and research organizations.

Finally, the authors thank Marilee Rowe, Kate Rudman, Kacey Christie, Eric Gray, and Aaron Ritter of Colorado State University (Fort Collins). Marilee Rowe served as administrative assistant for the project and largely organized the workshop arrangements. Kate Rudman and Kacey Christie were student assistants. Eric Gray attended the workshop and did an excellent job recording working group sessions. Aaron Ritter was the project's graduate research assistant and contributed substantially to the technical preparation of the report, as well as to the compilation of workshop results.

EXECUTIVE SUMMARY

CONTEXT OF WATER DISTRIBUTION CAPITAL MANAGEMENT

Water utilities use their distribution systems to deliver high-quality water to customers, in spite of breaks, corrosive deterioration, and other forces. These aging and invisible water distribution networks, spread over large areas with multiple connections and points of access, require large allocations of capital resources for renewal. The engineers and operating crews in utilities work in tandem with planners, information technologists, and finance staff to gather and utilize the data required to make effective decisions about maintenance and renewal. Often, these operations must be conducted in the midst of congested urban areas, without disrupting daily life.

This report is about the specific tasks involved in water distribution system capital management—locating, assessing, planning, repairing, and renewing water distribution systems. The report came about as a result of the Awwa Research Foundation's (AwwaRF's) project Synthesis Documents on Infrastructure Integrity (2001, RFP 2772), which aimed to help utilities assess the knowledge they need for distribution system capital management and to evaluate the available research-based tools. The report is a synthesis of knowledge and the suggestions of utility and other professionals about renewing distribution systems. In addition to utilities, the report is aimed at policy makers, and at the water supply research and development community (including AwwaRF), to help the water supply industry plan a responsive research agenda for distribution systems. Information presented in this report was gleaned from the project's literature survey, utility visits, surveys of experts, and three workshops. We also learned about practices in several different countries and in allied fields, including oil and gas and wastewater.

The report covers a broad range of technologies—from technology-based assessment tools for operators to capital strategy planning methods. The first topic, condition assessment (the "science" and engineering side of distribution system management), addresses locating assets, diagnosing failures, learning about deterioration mechanisms, and measuring condition. The second topic, planning and prioritization (the "management" side of distribution system capital management), requires the utility to find funding sources, relate to governing boards and the public, and provide management and financial accountability. The third topic (the "construction" side of distribution system capital management), comprises the physical work of maintenance, repair, and renewal.

Management technologies for these three areas are in flux. Some utilities believe that the replacement problem is urgent, requiring more national investment. Others believe that the problem is not serious and that pipes will last longer than projections show. Some utility representatives believe that they should rehabilitate or replace on the basis of observed failures, rather than try to assess the condition of pipe inventory. Some use advanced methods for planning and prioritization; others are only beginning to think about implementing formal methods. "One size fits all" does not apply to distribution system capital renewal methods.

In this report, an "issue tree" is used to show the components of the management problem and to provide a structure for analyzing problems. For example, one constraint for distribution system work is the different materials in the network and the lack of standard designs and configurations. In a similar way, solutions are illustrated with a tree that shows the three requirements for success—knowing what to do, being capable of doing it, and being motivated to do it. The requirements for success lead us to ask the questions of *who, what, why, how*, and *when*.

SURVEYS AND NEEDS STUDIES

This report reviews the surveys and needs studies that frame the agenda for distribution system renewal. Starting about 1980, these began to appear more often, after implementation of the Safe Drinking Water Act converged with emerging awareness of national infrastructure investment needs and with the ramp-up of AwwaRF programs. Attitudes changed over the past two decades. In the 1980s, utility managers seemed to think they were keeping up with their operations and maintenance (O&M) needs to maintain reliability. Then, studies began to show that large sums were required for repair, rehabilitation, and replacement work. However, the analytical base for the studies has not been consistent. In any case, today's replacement rate in the United States, about once in 200 years, does not appear adequate. Replacement value of distribution systems seems to be on the order of $1,500 per capita, and 20-year investment needs range from values of about $300 per capita up to values as much as four times higher. However, the credibility of the needs studies rests on the standards they reflect, estimates of condition, and service lives, and these need further research.

Before about 1960, short papers were published about problems with cast-iron water mains. To document needs, more data on distribution systems were needed. In 1983, the American Water Works Association (AWWA) surveyed 39 of the largest utilities, and later expanded the survey to create the Water Industry Database. This led to the development of Water:\Stats (AWWA 1997), which is available today. These data are useful in current projects for performance assessment and benchmarking. Generally, the surveys from the 1980s showed that awareness of distribution system problems was increasing and that there was significant variation in utility practices. Since then, many new materials, products, and technologies have been developed. The context to improve management of distribution systems is changing rapidly, and today, utilities have many tools at their disposal.

CONDITION ASSESSMENT

Of the new technologies, those available for condition assessment have emerged most rapidly. Ideally, condition assessment means that the condition of the organization's assets is continually being assessed so that return on investments can be maximized. From a practical standpoint, however, this is not fully realized because utilities practice varying degrees of maintenance and renewal strategies. These include "proactive," "reactive," "reliability-based," "on-condition," or "breakdown" maintenance and renewal. Regardless of the maintenance philosophy, our research suggests that more advances will be made in condition assessment than in other arenas of distribution system capital management.

The report offers a framework for condition assessment to serve as a common ground for information exchange. This is consistent with the findings in the field of benchmarking, where common definitions and measures are required. The framework adapts to system components and subsystems, to the technical and management sides of the organization, and to capital and operating components. It also links the parameters of condition assessment with the uses of the information for renewal work by operators, engineers, and maintenance personnel, as well as by budget and finance staff.

The capital indicator is infrastructure readiness or the question, "Will it last"? This indicator has two sources of information. One is how secure the infrastructure is from damage caused by natural or human-caused threats. The other is how robust it is against degradation, such as corrosion. The operational indicator is operations readiness or "Will it provide capacity"? This has two basic dimensions. One is that the component provides the quantity of water needed, and the other is that it supplies the quality of water needed.

The use of condition assessment information varies across organizational units. Whereas some believe that data ought to be collected by the unit using the information, it makes sense that condition information should also be shared among units. Then, these data can be used for operational evaluation of the need for preventive maintenance or repair; for engineering evaluations of the need for renewal, including rehabilitation or replacement; as budget information for the capital improvement program; and for accounting evaluation for Government Accounting Standards Board Statement No. 34 purposes.

Before condition assessment can be conducted, assets must be located and placed into inventories. New technologies for locating pipes are reviewed in this report, and the most recent AwwaRF report is summarized. Recommendations from an international workshop on distribution system asset location are also presented.

Pipe segments or distribution system components are assessed for structural condition, hydraulic characteristics, and for their influence on water quality. If the assessment of each component is valid, their aggregation should give a valid assessment of the network's condition. However, a network can also be assessed directly for its structural integrity, hydraulic characteristics, and water quality. Assessing components mainly involves structural tests; evaluating hydraulic capacity and water quality relate more to systemic assessment.

Two main technologies are used for destructive testing of components: coupon sampling and controlled destructive evaluation. A number of other technologies are included in another broad method, nondestructive evaluation (NDE). Coupon testing means to cut a sample of a pipe for visual inspection and analysis. It is a logical technique for old cast iron where corrosion or loss of wall thickness might be related to zonal soil conditions, pipe age, or related factors that could be generalized. However, spot condition assessment and coupon testing have definite limits and several disadvantages, such as cost, replicability, and disinfection. In controlled destructive evaluation, the utility can find out directly if a reach of pipe has integrity. Experience of a large-scale program by the Los Angeles Department of Water and Power is cited.

NDE technologies hold great promise, but their effectiveness and cost vary considerably. Three variables frame the possibilities: type of pipe, technology used, and information sought.

Early NDE involved inspections and use of techniques such as sensing the quality of a metal by its ring after tapping it with a hammer. Techniques using electromagnetic current induction and magnetic field detection were developed. Gamma radiography, penetrant testing, eddy current testing, and ultrasonic methods came along in the 1930s. The Society for Industrial Radium and X-Ray Testing became the American Society for Nondestructive Testing, and new technologies, such as radar, acoustic emission, and thermal imagery, came along in the 1940s and 1950s. Today, reports on NDE describe techniques using these and other technologies for many industrial and military applications.

The main technologies used today are magnetic, electromagnetic, sonic, acoustic, infrared thermography, and ground-penetrating radar equipment for locating pipes; global positioning system/geographic information system (GIS) databases for managing information; and ultrasonics, acoustic emission, magnetic flux leakage, and remote field eddy current (RFEC) for assessing metallic pipe. Other technologies, such as closed-circuit television, have more limited applications.

Methods for assessing distribution systems, as opposed to components, focus on hydraulic analysis, water loss accounting, and use of pigs. In each of these categories, advances enable utilities to mount effective assessment programs, depending on the circumstances. Hydraulic and water loss analysis are quite advanced, but use of pigs is limited. When technologies are developed that can send pigs around sharp bends and through valves, their use may increase.

In addition to assessing condition, the utility must determine what condition implies for predicting time to failure. This requires understanding of failure mechanics and advanced utility data management systems to keep up with failures and investigations in the utility.

PLANNING AND PRIORITIZATION

Utility tasks in planning and prioritization span a spectrum from problem assessment to making decisions about solutions. This spectrum is embodied in the capital planning, programming, and budget process and is described in the *Capital Strategy Planning Manual* (AwwaRF, CH2M HILL, and AWWA 2002). This report focuses on the prioritization process itself, as it fits into the capital planning process.

In prioritization, utilities face the dual problem of identifying their highest priority pipe renewal projects and comparing them to other capital investments, which are often more visible and attractive. These start by rating alternative pipe renewal projects according to how they score in an evaluation process with rules and criteria and reduce to comparing and justifying pipe renewal projects in the capital planning process.

The criteria—such as reducing failure frequency and improving water quality—are selected to help achieve utility goals in areas such as customer service and regulatory compliance. The challenge is for the utility to ascertain, for individual pipes, the fragility function, or time to failure and vulnerability under varying scenarios.

The most direct way to evaluate fragility is with condition assessment technologies and direct inspection techniques. However, the sheer magnitude of buried pipe and the difficulty in accessing pipe make this problematic. This area needs more research and product development.

Another approach to determining fragility is to lump pipes into classes and assign to them average failure information. That way, with statistics of an overall system, the utility can predict needed investment rates to achieve a certain renewal target. This approach is used in the KANEW (modeling software used to predict future replacement needs) and Nessie models.

Regardless of the approach, the utility should have the capability for continuous learning about its distribution system. The water main condition database is a hidden weapon for this, and it can be included in a commercially available "enterprise" software package that is linked to a GIS package, or it can be created and maintained by the utility, or both. In its water main condition database, the utility should collect, little by little, all the bits and pieces of information required to know the condition of water mains for renewal decisions.

The water main condition database will normally be linked to the asset inventory. In other words, information such as age, type of material, and location are naturally associated with information items such as repair history and inspection results. Clearly, the amount of information to be associated with a given pipe asset is formidable and can be used for several purposes. Managing and utilizing all these data requires excellent technical skills, so we recommend that, along with a water main database, the utility maintain a highly skilled analysis staff.

Conducting the decision analysis is the other component of prioritization. Decision analysis means taking all information related to a particular decision, delivering it to one place, arraying it, evaluating it, adding value to it, and using it to come up with an indication of how to proceed. This task has received much attention in the research literature, and if the fragility of individual pipes is known, utilities are fully prepared to set priorities for renewal.

The earliest tools for decision analysis described in the research literature are based on economic approaches. These offer analysis tools like "break-even" analysis, "rate-of-return" analysis, and so forth. These can be useful, but they tell only part of the story, in that they cannot compare all of the operational, social, and other factors that the utility deals with. These factors are summarized later in this report.

The latest comprehensive tool in an AwwaRF report is the "decision support system." These are integrated software packages that enable utilities to assess problems and plan solutions within a single software package that also includes data and help resources for the analyst.

RENEWAL METHODS

In general, repair, rehabilitation, and replacement practices among utilities vary greatly. All utilities repair broken pipes and leaks, but some utilities rehabilitate pipes regularly. Others focus on replacement without much rehabilitation activity. At the project workshop, participants observed that North American water utilities are reluctant to use newer technologies for pipe renewal. Reasons are lack of awareness, cost effectiveness, liability, and uncertainty about sustainability. Options mentioned to speed adoption of new technologies were cost reduction by assembling larger projects; offering training; providing education for management, politicians, and customers about benefits of new technologies; and using the private sector for construction of renewals.

The lines between repair, rehabilitation, and replacement can be blurred. To give a progression from repair to replacement, definitions are presented, and a task map is included to classify work by processes. The task map can be used to discuss jobs, tasks, and learning requirements, and it shows how the workforce can be cross-trained to carry out similar tasks that involve different processes and technologies.

The framework for organizing maintenance is the "maintenance management system," which takes a systems approach to maintenance. A recent AwwaRF report (Deb et al. 2000) concluded that maintenance procedures emphasize flow and pressure but neglect reliability and water quality. The report recommends measures for enhancing reliability and water quality, improving O&M, enhancing performance, and managing information. Utilities are also studying concepts such as "reliability-centered maintenance." Techniques such as this may lower cost and place a focus on critical equipment and components, which are defined by water quality violations, productivity, regulations, and safety. The techniques are data-intensive and require a multidisciplinary team, including maintenance, mechanical, electrical, and instrument engineers, as well as operations personnel.

Repairing a leak or break requires locating the problem, getting organized, and completing the repair tasks. New technologies and methods are emerging for locating problems, using better equipment for excavation, and applying materials to repair the problem. These include more accurate location, less-invasive excavation, and new approaches and materials for repair. Trenchless methods for renewal have advanced in the last two decades, but they have not made the inroads in water distribution systems that they have in sewer collection systems. Water systems have much more hardware, operate under pressure, and are more difficult to access. AwwaRF published two reports on trenchless rehabilitation and replacement (Deb et al. 1990; Deb, Hasit, and Norris 1999). The latest report discusses pipe-bursting methods (hydraulic, pneumatic, and static pull) and lining methods (rolldown, sliplining, and cured-in-place pipe) for replacing water mains, documenting the applications on video. Findings include recommendations on appropriate technologies, which can be presented in the form of a decision tree, and information on problems encountered during demonstration projects. AwwaRF has also published several reports on maintenance and pipe-cleaning techniques.

For rehabilitation, utility representatives can choose from a family of nonstructural or semistructural lining techniques, or they can replace a pipe with a new structural liner or pipe bursting, followed by insertion of new pipe. The oldest technique is cement–mortar lining, which has been used for 50 years. Epoxy lining can also be used and has been adopted in Europe more than in the United States. AwwaRF has supported several projects to evaluate epoxy lining. Cement–mortar lining has advocates and detractors, and the promise of quick-setting epoxy—perhaps even with some structural qualities—is attractive. Sliplining can be used inside old pipes, but at the cost of losses in diameter and asset capacity.

The daunting aspect of the renewal problem is the difficulty in accessing distribution systems. The problem is much more complex than in other pipe scenarios because of congestion, pressurization, hidden assets, age, water quality, and service connections. Despite these barriers, innovation in renewal methods has proceeded apace. Utility managers now have many more methods to choose from than in the past and, in many ways, are now limited more by economics than by technology.

The private sector is working to develop new, commercially attractive technologies. Research suggested by participants includes documenting case histories, evaluating technologies, analyzing barriers to the new technologies, and reviewing European practices.

CORROSION AND CORROSION PROTECTION

Given the many miles of metallic pipe and the many environmental conditions that must be faced, corrosion is and will continue to be a big issue for utilities. Corrosion is costly, with costs to utilities of about $48 billion per year, according to one author (Jeyapalan 2000). It is a complex subject with many facets, and research has produced significant findings about it. However, because of its complexity and the many situations that can arise, utilities are not going to discover dramatic and easy solutions to corrosion.

As a result of studies by AwwaRF and others, a great deal is known about internal corrosion, but it remains a very complex process. One set of authors reviewed almost 300 articles that embrace 100 years of water utility experience with corrosion (McNeill and Edwards 2001). They studied factors including water quality and composition, disinfectant residuals, pipe age, scale formation, temperature, flow conditions, biological activity, and corrosion inhibitors. The authors also reviewed how proposed regulations may affect iron pipe corrosion. They concluded that iron pipe corrosion is affected by practically every physical, chemical, and biological parameter in water distribution systems. In general, water supply industry professionals believe that the industry should modernize with regard to internal corrosion control. An international team of authors concluded that utilities have not kept pace with corrosion science and have only recently begun to import corrosion technologies from other industries (AwwaRF and DVGW-Technologiezentrum Wasser 1996). They recommended standardized coupon procedures to yield data on corrosion, surface scales, and pitting. These authors thought that a database on materials corrosion and water quality should be organized and that standards for electrochemical measures are needed.

Perhaps less is known about external corrosion than about internal corrosion. External corrosion can be affected by soil resistivity, pH, and other factors. Utilities should correlate soil and external pipe corrosion data to learn about their unique situations. Research advances using GIS show promise for tracking the data. One study offered guidance about optimum time to replace cast-iron water mains after considering corrosion. The most promising technologies were ultrasound and RFEC. Ultrasonic inspection is simpler but will not work with tuberculation, so RFEC is more flexible.

Cathodic protection retrofits offer promise for extending the life of metallic pipe. Although it seems to be cost effective, our workshops and the literature showed a variety of opinions. Some utilities are convinced that it works and is cost effective; others are not sure.

Corrosion and its control will continue to be an important research subject for the future. Reports we reviewed presented numerous individual topics that are summarized later in this report.

WATER QUALITY

Although water quality is a main area of concern for distribution systems, it was not a focus area in this project, and the report reviews the main issues only briefly. Most experts believe that water quality will be perhaps the biggest driver of distribution system renewal in the future. They think that many unanswered questions remain about what happens to water quality after water enters the distribution system. To detect changes in water quality, monitoring can be done for regular operations or to detect intrusions. Water quality modeling is also a promising technique, but it is not currently practiced by many utilities. Areas of research on water quality that require attention include monitoring, modeling, distribution system design and renewal, and distribution system management to protect water quality.

UTILITY PROGRAM FOR DISTRIBUTION CAPITAL MANAGEMENT

This report includes a chapter that summarizes the tools available to utilities to plan and manage their distribution capital program. The basic tool kit consists of the following:

- An inventory as the core activity
- A condition assessment
- A maintenance program (preventive and corrective maintenance)
- A system planning and needs assessment
- A financial accounting of the system
- An asset management framework
- A capital improvement plan and program
- A capital budget
- A capital management system (strategy and framework)

Each of these tools has a research base, which is explained briefly in the report.

DISTRIBUTION SYSTEM ISSUES

Issues that remain of concern include the high levels of projected necessary investment, along with required improvements in the credibility of needs studies. Utility management must also continue to develop all categories of the utility's workforce to manage distribution systems. In planning, prioritization, and performance evaluation, improved methods continue to be an issue, but uncertainty about expected service lives of water mains makes it impossible to estimate investment needs precisely. The need for research about pipe materials, structural characteristics, and construction practices has continually been cited at workshops and during projects and this need will continue (particularly in the area of better understanding failure mechanics). Condition assessment shows great promise for distribution systems management, but more practical methods for use by all utilities are required. Although high-quality water is required at the tap, the shift in responsibility at service connections creates unknowns and uncertainties about quality control. Data management has emerged as one of the principal issues in distribution systems management.

RESEARCH NEEDS

Research needs identified during the project generally tracked the issue areas described later in the report. Some of the strategic goals mentioned are

- To make progress on institutional issues to promote innovation
- To design a policy for public investment
- To overcome barriers to new technologies
- To provide an improved regulatory framework for buried infrastructure management
- To stimulate research and testing on utility databases for pipe management and develop a national database of failures, causes, and renewal methods, including standard reporting methods
- To continue research on failure mechanisms with different types of pipe and causes for pipe, joint, lining, and coating deterioration
- To codify design information
- To create more effective methods for installation of service connections

Although AwwaRF has sponsored a number of projects on distribution systems, additional individual projects are required. Examples of projects needed include new and, in some cases, continuing research in the following areas:

- Promoting innovation and public investment, and overcoming barriers to new technologies
- Funding mechanisms and procedures for improving rates of renewal and improvement
- Technology transfer and workforce development
- Organizational systems for handling the large amount of data and analysis required
- Utility databases for tracking distribution system needs
- Ways to determine which pipes should be tested or replaced
- Prediction of service life
- Pipe materials and structural qualities

- Development of a national database of failures
- Methods for inspecting and assessing all types of pipes
- Development of instrument systems
- Development of improved pigs for water distribution systems
- Codification of design information
- Design of distribution system performance standards and lining techniques for rapid return to service
- Corrosion measurement and control, water quality maintenance, and regulatory issues

If any single finding stands out from this work, it is that a process is needed to periodically bring together research findings in focused areas and publish the condensed findings in ways that are helpful to utilities. At the same time, AwwaRF would continue its present programs for identifying needs and carrying out applied research until the findings are delivered to the users. In the case of distribution systems, the area of condition assessment stands out as most in need of further classification, leading to publication of condensed reports in applied areas.

CHAPTER 1
INTRODUCTION TO DISTRIBUTION SYSTEM INTEGRITY

PURPOSE OF THE REPORT

Nationally, the integrity of about 1 million miles of water mains is critical to the health and welfare, as well as to the finances and operating performance, of water utilities. Although these water mains are mostly invisible, their replacement value is one-half to two-thirds of the asset value of most utilities.

This report has been prepared to help utilities evaluate the research-based tools available to manage these water mains as capital assets. Distribution system issues have been studied for many years, but with mounting investment needs and concerns about system performance, the state of knowledge requires review. This report contains a synthesis of knowledge and the suggestions of utility and other professionals about renewing distribution systems. Three specific topics are covered: how to assess condition of systems; how to set priorities for renewal; and how to perform repair, rehabilitation, and replacement. In addition to utilities, the report is aimed at policy makers, and at the water supply research and development (R&D) community, including the Awwa Research Foundation (AwwaRF), to help these professionals plan a responsive research agenda for distribution systems. Each topic is introduced with background knowledge, and the research base is reviewed. Utilities can use the report as a guide, and the water supply industry can use it to plan future research.

INFRASTRUCTURE INVESTMENT NEEDS

Building and renewing distribution systems requires massive investments, and today's U.S. replacement rate, about once in 200 years, does not appear adequate. However, the needs estimates require clarification.

Water supply industry representatives are unsure whether utilities should continue to renew at the current rate or to raise rates to accelerate renewal, or whether government subsidies should pay for health and safety objectives that go beyond pure services. This uncertainty has been reflected in policy studies that project large sums required to repair, rehabilitate, and replace distribution systems, as well as in the risk aversion that characterizes utility decision making about pipe renewal. However, the credibility of the needs studies rests on the standards they reflect, estimates of condition, and service lives.

The needs studies increased after about 1980, after the Safe Drinking Water Act (SDWA) was passed in 1974 and the media-driven infrastructure "crisis" emerged. This period provides a starting point for reviewing the policy studies. In 1980 the Comptroller General issued a policy study on distribution systems. Then, in 1986, O'Day et al. wrote that there were divergent views on needs for upgrading distribution systems. Although these divergent views stemmed from lack of research and standard procedures to assess condition, most utilities seemed to think that they were keeping up with their operations and maintenance (O&M) needs to maintain reliability. During this period, the water supply industry started to see the need for data, leading to programs designed by the American Water Works Association (AWWA) and the U.S. Environmental Protection Agency (USEPA) to maintain water industry statistics.

Even though the evidence of need to improve distribution system infrastructure has continued to build, it still requires more rigorous and scientific analysis. The lack of rigor in proving needs is still based on the issue noted by O'Day et al. (1986)—lack of standards and measurement procedures.

The procedure for a needs study starts with the inventory and uses condition indicators to assess the stock of pipes. For example, one indicator might be the pipe age and another might be the roughness factor. Condition should be compared to a standard to determine the need for improvement, which would be extended over a selected time horizon to develop a needs assessment. Procedures for condition assessment are discussed in chapter 5.

After O'Day's report, a sequence of studies showed that consistently large sums were required for repair, rehabilitation, and replacement work, but the analytical base for the estimates was not consistent. The National Council on Public Works Improvement (1988) report card gave water supply a "B–" and complimented the sector as an effective, locally operated service that would need drastic rate increases in the decade after the 1986 SDWA amendments. The report card also expressed concern about pricing below cost of service, increasing cost to meet purity standards, and deteriorating distribution systems in older cities. This anecdotal evidence of need for investment was based on focus group opinion rather than rigorous analysis.

More than a decade later, the American Society of Civil Engineers (ASCE 2001) downgraded the drinking water grade to a "D" and cited an annual shortfall of $11 billion needed to replace facilities nearing the end of their useful life and to comply with federal water regulations.

In a 1999 needs survey required by the 1996 SDWA amendments, USEPA (2001) found that $139 billion was needed over the next 20 years (to 2018) to meet SDWA regulations and ensure delivery of safe water. Of the $139 billion, $81.9 billion (or 59%) was for transmission and distribution (T&D) systems.

USEPA's survey required utilities to submit projects, and USEPA evaluated them against established criteria. These "needs," then, are based on utility perceptions of projects required over 20 years to protect public health, including preserving physical integrity of systems, conveying treated water to homes, and ensuring compliance with SDWA regulations. No new growth is included because it would not be eligible for funding under the revolving fund program.

Meanwhile, AWWA had commissioned a study by Stratus Consulting (1998), which conducted an independent assessment of distribution system needs that led to a 20-year estimate of $325 billion, or roughly 3.9 times the USEPA estimate. In 2001, AWWA surveyed 20 utilities of varying locations and sizes, and reported needs of more than $250 billion over the next 30 years to replace pipes and infrastructure (*ENR* 2001). This does not include more than $12 billion per year that utilities already spend on infrastructure repairs or SDWA compliance.

The Water Infrastructure Network (WIN 2001) issued a report in 2001 asking for a 5-year, $57-billion federal investment in drinking water, sewer, and stormwater infrastructure. These estimates are mostly based on utility reports of their needs, or in the case of the 2001 AWWA study, they are based on assigning a lifetime to pipes, where in fact the actual lifetimes are not known. Needs estimates build on each other to some extent; that is, perceptions of the need may increase by a shared understanding of the problem. In any event, the finding by O'Day et al. (1986) still holds—standards are needed against which to evaluate the needs.

Although standards for assessing needs are required, the complexity of distribution systems makes them very difficult to develop. In any event, developing national needs assessments is an inexact science. From the estimates, however, we gain an appreciation of the magnitude of the problem. It is also true that perceptions vary among utilities. Some utility managers believe that the replacement problem is urgent, requiring more investment. Others believe that the problem is blown out of proportion and that pipes will last longer than the projections. As the debate continues, the water supply industry can do much to improve its systems by applying research-based technologies.

WATER DISTRIBUTION—A MERGER OF OLD AND NEW TECHNOLOGY

This project covers a broad range of technologies—from technology-based assessment tools for operators to capital strategy planning methods for planning staff. For that reason, an overview of the general situation may help to provide perspective. In general, an old technology—buried water pipes—is being managed with increasing sophistication. Old pipes remain in place, but new methods for assessment and renewal are in use.

Condition assessment is not uniformly appreciated, and some utilities believe in a reactive approach, where they rehabilitate or replace pipes on the basis of failures rather than try to assess condition of pipe inventory. Practical issues, such as replacing pipe when streets are renewed, may also govern decisions.

Some utilities use advanced methods for planning and prioritization; others are only beginning to think about them. Research studies offer new methods such as KANEW (a modeling software used to predict future replacement needs), the Nessie curve, and optimization, but use of these methods is at a very early stage. In fact, judgment-based decision making is the predominant method, with rehabilitation and replacement usually taking place only after failures become pressing.

The status of renewal technologies is also advancing rapidly. Utilities can choose from more materials and newer methods than ever before. New technologies must still be validated, and economics and institutional barriers are important issues.

Although the field of pipeline engineering involves other sectors—such as oil and gas—water pipelines are different because they serve congested urban areas and involve more hardware (fittings, local services, loops, and diverse types of pipe, for example) than oil and gas. Urban systems are simply more complex. Some technologies from oil and gas can also be used for water supply lines, but opportunities for such transfers are limited. Renewal in water distribution systems is also more complex than in wastewater systems, and even though new and innovative methods are coming online, methods vary widely among utilities.

MANAGEMENT TASKS AND CAPITAL LIFE CYCLES

In recent years, many new concepts and terms for management have emerged. To place these into perspective, two diagrams are presented. Figure 1.1 is a "task map" for capital management. Note the interdependence of the work activities that include maintenance, planning, accounting, and engineering. Using the basic capital inventory, these produce condition assessments; the needs assessment; the maintenance management system; the system accounts, including Government Accounting Standards Board Statement No. 34 (GASB 34) information; the asset management system; and the capital improvement plan and budget (Grigg 2002). Also note that information compiled for condition assessment and planning is used for many different purposes.

In a similar way, the infrastructure life-cycle (Figure 1.2) shows when different tasks come into play. Both new growth and renewal are part of the cycle, as well as engineering and construction, O&M, and condition assessment.

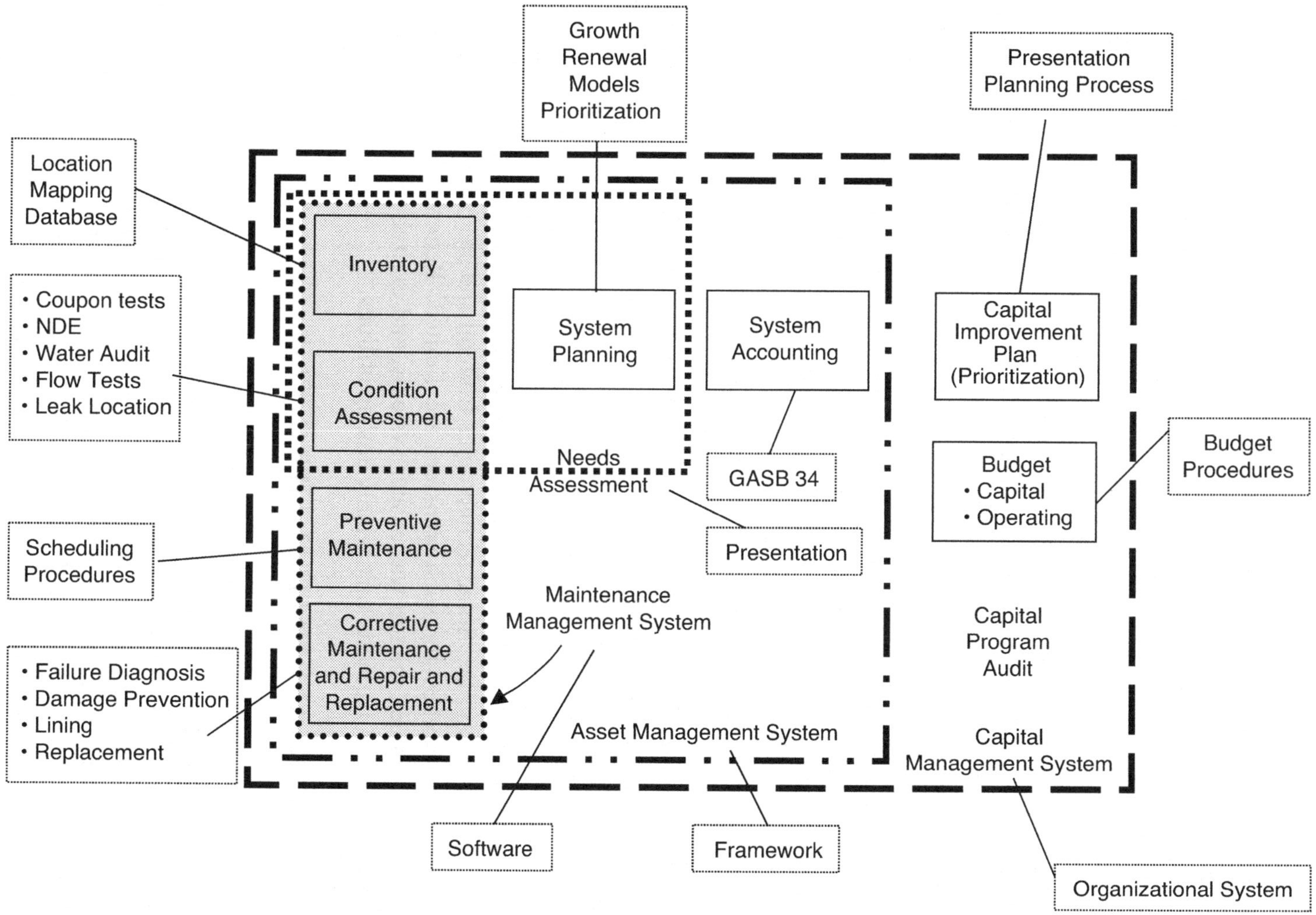

Figure 1.1 Capital management task map for water infrastructure integrity

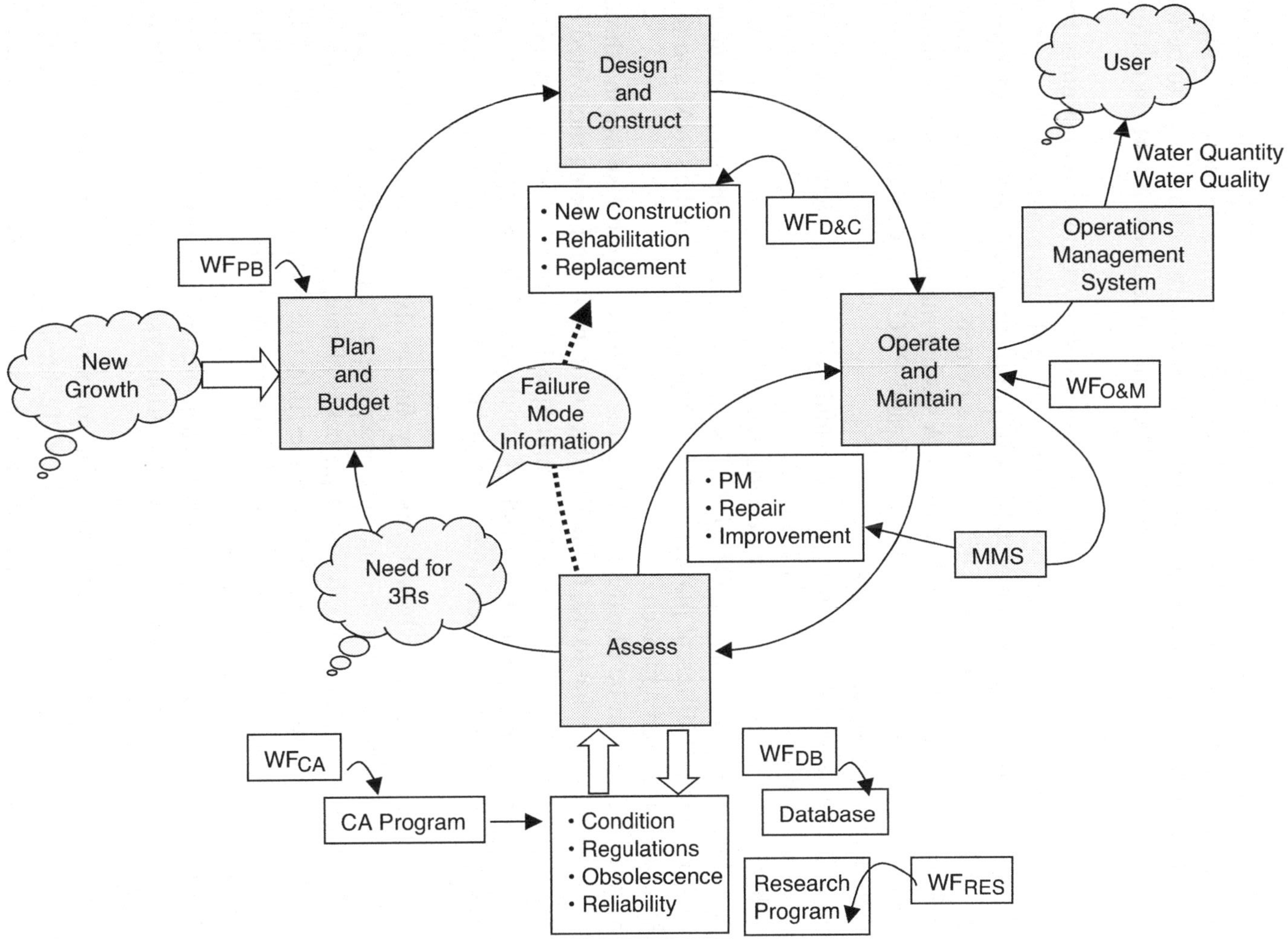

CA = Condition Assessment
DB = Database
D&C = Design and Construction

MMS = Maintenance Management System
PB = Planning/Budgeting
PM = Preventive Maintenance

RES = Research
3Rs = Repair, Rehabilitation, and Replacement
WF = Workforce

Figure 1.2 Life-cycle management for water distribution system management

IDENTIFYING THE PROBLEM AND THE SOLUTIONS

In any problem-solving exercise, the first requirement is to identify the problem. We address this with an "issue tree," shown for distribution system integrity in Figure 1.3. Note that the four levels of the issue tree also establish performance variables that can be measured.

The problem divides into capital and operating issues, then into branches of indicators. These yield ten indicators of distribution system effectiveness that are similar to those developed by Deb, Hasit, and Grablutz (1995). Each indicator measures a variable that either leads to better distribution systems or shows that it needs improvement. Deb and colleagues' report on performance indicators is summarized later.

The issue tree furnishes a structure for problem analysis. For example, constraints for distribution system work include the different types of material in the network and the lack of standard designs and configurations. This is an inevitable result of the original cast-iron networks, which have materials such as ductile-iron pipe (DIP), asbestos cement (A–C), plastic (polyvinyl chloride, or PVC), and polyethylene (PE) spliced into them. A system might include, for example, original cast iron, newer ductile iron, and sections of A–C and PVC. Crews require tools and training for the different kinds of pipe and fittings.

Another constraint is that the crews that find breaks and perform emergency repairs may be different from the crews that assess the condition of the pipe. Because condition assessment uses advanced technologies and databases, the engineering section may handle them, rather than O&M staff (although it should be noted that the O&M staff is also becoming more sophisticated). Similarly, the planners may include different players, but they must work with the O&M and the engineering staff, taking an integrated approach.

In a similar way, solutions can be illustrated with a tree (Figure 1.4) that shows the three requirements for success—knowing what to do, being capable of doing it, and being motivated to do it. The elements of the figure can be reduced to the simple questions of *who*, *what*, *why*, *how*, and *when*. These divide into categories of solutions, each of which will yield one or more management practices and assessment variables. Taken together, they involve the whole organization in one way or another.

In studying solutions, we illustrate how technology has evolved from early forms. A perspective on the issue can be gleaned from "technology assessment," a technique that evaluates technologies for their suitability for achieving different purposes (Braun 1998). For purposes of this report, we divided water distribution system technologies into five branches: pipe materials; pipe engineering; pipe construction, including repair, rehabilitation, and replacement; assessment and modeling; and management and prioritization.

As Figure 1.5 shows, technologies and methods in each area have evolved, and the total effect leads to a sharply increasing technology base for distribution systems. The various methods are explained in subsequent chapters.

Some technologies may seem more appropriate for cooperative, government-sponsored development, but the private sector participates in all of them. Technology for planning, condition assessment, and renewal work can be proprietary, and if a company offers an instrument or method on a proprietary basis, it is titled to recover its development expense through charges for its use. By the same token, government and the water supply industry develop technologies for condition assessment, renewal, and planning. Public investment may speed the development cycle, and the private sector can adapt the technology for its purposes.

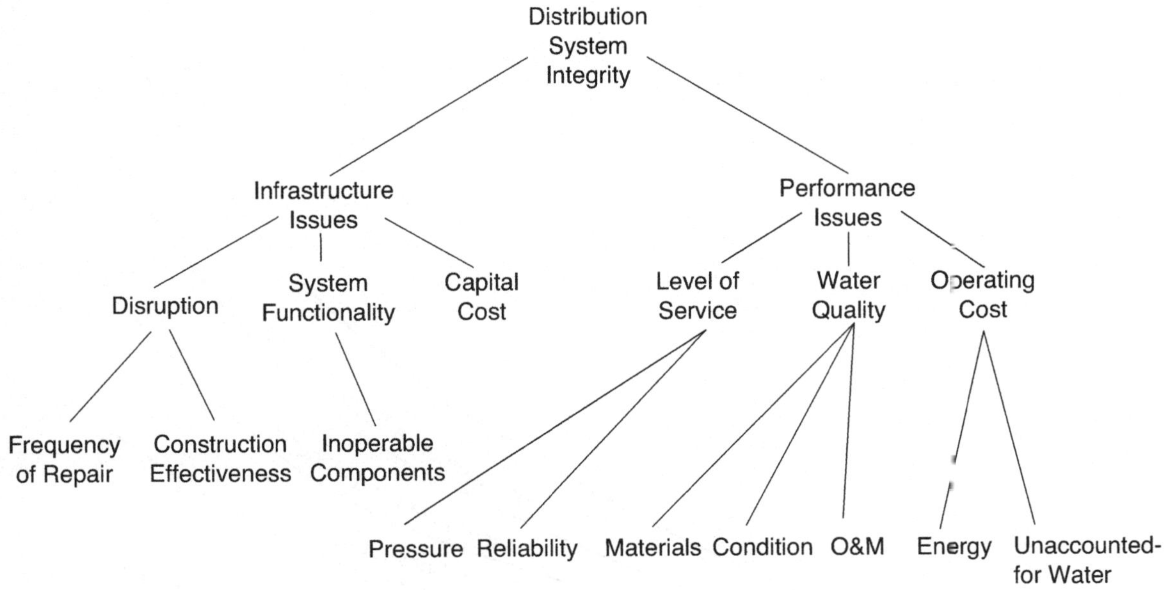

Figure 1.3 Issue tree of distribution system integrity problems

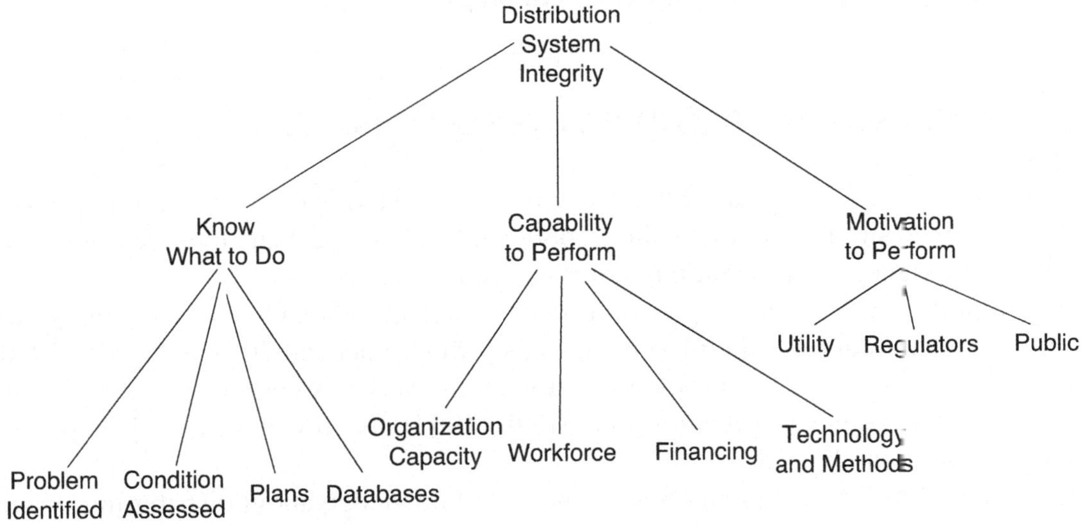

Figure 1.4 Issue tree of solutions to distribution system integrity problems

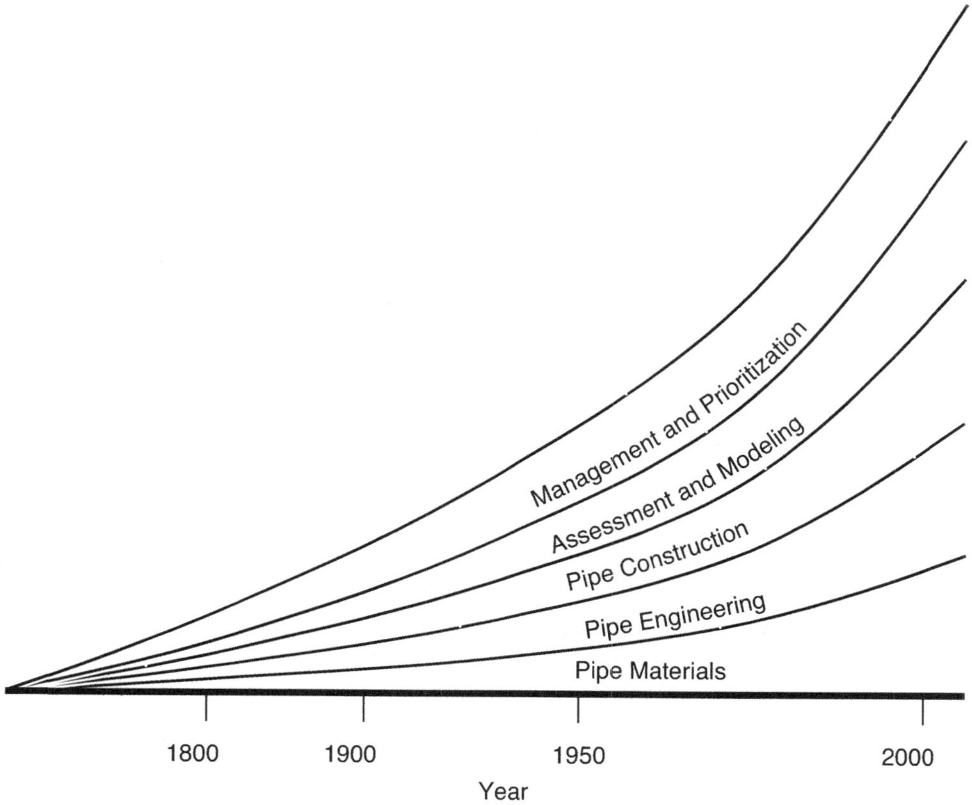

Figure 1.5 Growth of distribution system technology

GENERAL GUIDES ABOUT DISTRIBUTION SYSTEMS

A number of general guides have recently been written about distribution systems (cited below). They cover similar topics to this report, and, taken together, they contain substantial knowledge about managing distribution systems as capital assets.

The report by O'Day et al. (1986) presents excellent data and analysis about water main evaluations for rehabilitation and replacement as they were practiced 20 years ago. In 1990, Male and Walski wrote an overview of problem identification, causes, solutions, and decision making. The 1994 report by Kirmeyer, Richards, and Smith (1994) also reviews the full scope of issues, with emphasis on research needs.

Ellison's (2001) report presents a synthesis of knowledge about distribution systems. He covered a number of key topics, such as obtaining and managing data, evaluating materials and performance, studying corrosion, determining the condition of the system, assessing condition of individual pipes, implementing pipeline renewal technologies, and managing a pipeline renewal program.

Hughes (2002) presents a general guide with state-of-the-art articles on a very wide range of articles covering most of the pertinent topics relating to distribution systems integrity:

- The infrastructure gap
- Asset management

- Infrastructure management systems and software systems
- European models and databases
- Materials engineering
- Cast-iron pipe (CIP) failures
- Water main renewal planning
- Condition assessment
- East Bay Municipal Utility District's (EBMUD) replacement program
- CIP management by cathodic protection and hydroscope testing
- Statistical prediction of water main breaks
- Financial and economic optimization
- Financial modeling
- Revolving fund approaches
- Economic approaches to renewal
- Overall integrity management
- Planning
- Trenchless methods
- Pipe bursting
- Performance criteria
- Prioritization
- Renewal contracts management
- Cement–mortar programs
- Public involvement
- Insurance to set priorities
- GASB 34

In addition to these, AWWA published a text (AWWA 2003) and McGraw-Hill published a handbook on distribution systems (Mays 2002). Also, a number of conferences about pipeline issues have produced proceedings, and other fields, such as the oil and gas industry, have published research results. In Guerrero's (2002) testimony, for example, he explains how lax safety programs have led the Office of Pipeline Safety to implement an "integrity management" program for the safety of some 2.2 million mi of pipelines, mainly oil and gas. This is a risk-based regulatory approach. The Office of Pipeline Safety oversees three types of pipes: about 322,000 mi of natural gas transmission lines; 1.7 million mi of natural gas distribution lines; and about 155,000 mi of hazardous liquid lines, including crude oil and refined products such as gasoline. Industry resources, such as the Pipeline Research Council International (PRCI 2002), have compiled research results. This publication, which is available for sale, included review of 14 references describing pipeline failures and testing methods. The council has a great deal of information on management of steel pipelines and on corrosion control.

WATER UTILITY DECISIONS

Distribution systems present an important management issue to water utilities, regulators, and the financial community. Managing them requires substantial data and sophisticated analysis. The general situation in the industry is that the problem presents both capital and operating issues, which require solutions that involve a range of roles in utilities.

Overall, utilities face a decision tree somewhat like the one shown in Figure 1.6. First, utility managers must determine if distribution systems require significant attention, and if they do, they turn to three questions: what is the condition; what are the priorities; and what methods should they use for assessment, planning, and renewal? Each of these questions leads to further questions and decisions, which are discussed later in the report.

We present the technologies in the form of decision trees, which involve branching decisions, where one decision leads to a series of choices. In the management literature, various forms for decision trees are given. In this report, a decision tree is basically an array of technology choices. Because these can become quite detailed, the approach taken is to give an overall decision tree in this chapter (Figure 1.6) and to present the detailed decision trees by technology in subsequent chapters.

SOURCES OF PROJECT INFORMATION

Information presented in this report was gleaned from the project's literature survey, utility visits, surveys of experts, and three workshops. Two regional workshops were held, and the project's Distribution System Infrastructure Integrity Workshop (referred to as "the project workshop" in the rest of this report to differentiate it from other workshops mentioned) was held on April 3–4, 2003, at Colorado State University in Fort Collins. Its purpose was to exchange experiences about distribution system integrity; report best practices for condition assessment, planning and prioritization, and system renewal; and provide an industry information source on distribution system integrity. More than 40 participants, including representatives from U.S. and international utilities and other consultants and experts, gathered to exchange knowledge and lessons learned.

Participants generally thought that aging and deterioration of distribution systems is a serious issue. They do not think, however, that the problem is as simple or one-dimensional as the aging of the infrastructure, and some think that, even though it is serious, the issue has been overemphasized. On the other hand, others think the emphasis is necessary, drawing to it attention that is producing better management practices.

In general, participants agreed that the nation needs to spend more on distribution system renewal. However, a key point emerged—that *utilities must continue to get better information on condition, replacement rates, and maintenance needs*. Utilities need better technology and management methods and to get smarter, not necessarily spend more.

At the beginning of the project workshop, utility representatives noted that the replacement problem is urgent, requiring more national investment. Others believe that the problem has been blown out of proportion and that pipes will last longer than projections show.

Preliminary results suggested that condition assessment is controversial, as some utility managers believe that they should simply rehabilitate or replace on the basis of observed failures, rather than try to assess the condition of pipe inventory. The project workshop participants, who may tend to be believers in technological methods, showed more confidence in science-based condition assessment than many other utility employees.

Some utilities have advanced methods in place for planning and prioritization; others are only beginning to think about implementing formal methods. Studies have been completed to offer new research-based methods. The project workshop showed that "one size fits all" does not apply. Methods are available, but they need to be presented better. In particular, a method that can be used for small utilities is lacking.

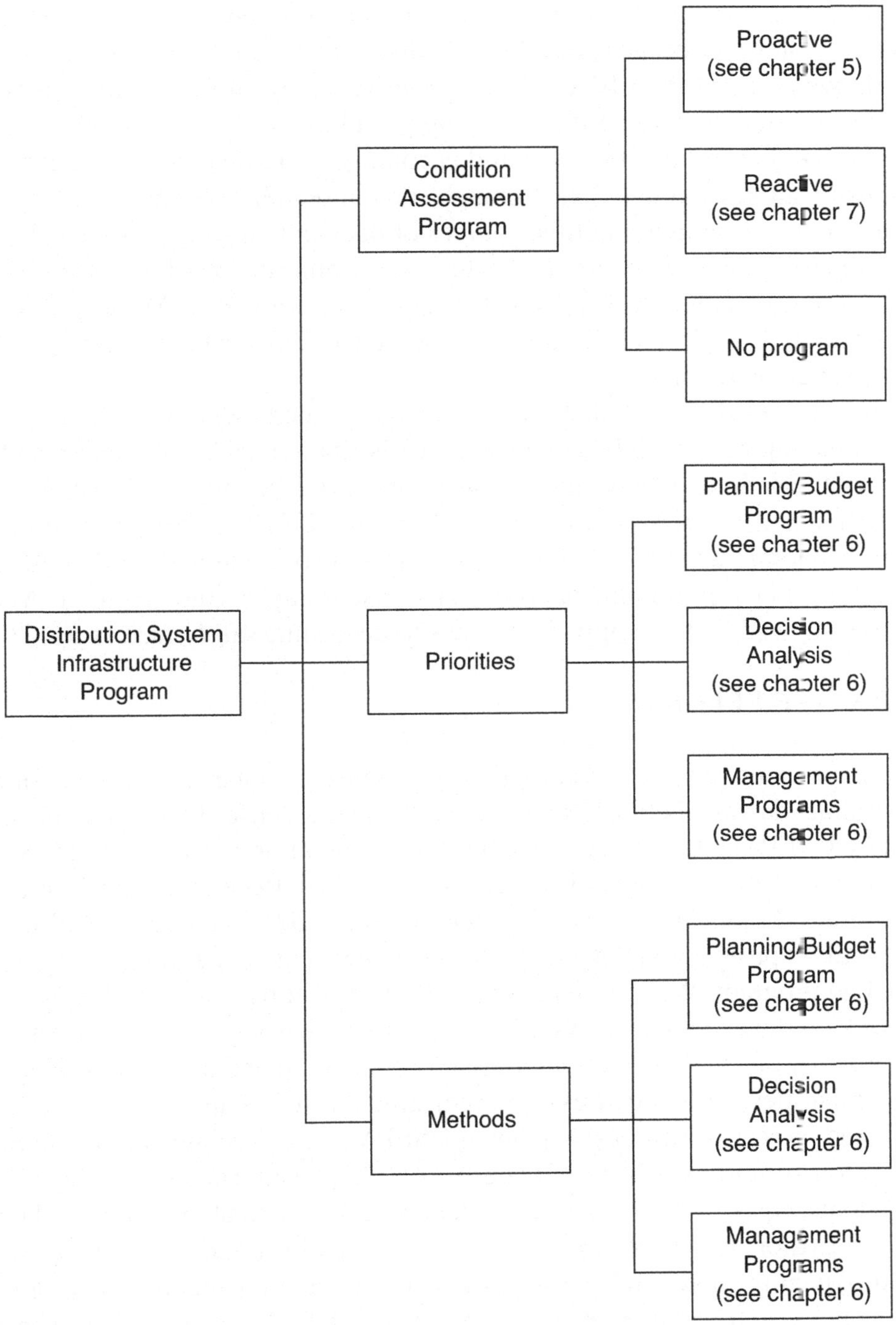

Figure 1.6 Decision tree for infrastructure program

The project workshop confirmed that renewal in water distribution systems is more complex than in wastewater because of all the hardware involved, but a number of new and innovative methods were reported to be coming online. Even so, there is wide variation in use of methods among utilities, and some utilities simply replace rather than rehabilitate. The project workshop suggested that further studies would encourage more innovation and experimentation.

On prioritization, we found that, although some research has been done, sophisticated utilities are finding ways to compile multiple sources of data and knowledge to develop multifaceted capital improvement plans. For example, Sydney, Australia; the Washington Suburban Sanitary Commission (WSSC; Laurel, Md.); New Orleans, La.; East Bay Municipal Utility District (EBMUD; Oakland, Calif.); and Columbus, Ga.; cited different examples of this approach (summarized later in the report).

The project workshop included presentations by experts who have done AwwaRF studies themselves. For example, Deb (2003) reviewed the distribution system situation and showed how the older parts of the nation still have more cast-iron pipe (CIP), while the West has more A–C and less DIP. He explained the KANEW model and the break-forecasting procedure, then gave an example of an economic analysis of replacement. Deb also illustrated models for pipe loading, pipe deterioration, and pipe failure analysis. His case study of the Regional Municipality of Ottawa-Carleton showed how to apply the models to determine safety factors for CIP.

COLLABORATIVE EFFORTS

Because competitive issues and proprietary systems are not a central issue for most technologies, collaborating on distribution system research makes sense. Collaborators include federal agencies, mainly USEPA, other research institutes, and similar organizations in other countries. For this project, literature from a number of sources was surveyed. Recently, Zhang (2003) summarized AwwaRF's program; Pramanik, Blankenship, and Moeller (2003) covered the Water Environment Federation's (WEF's) program; and Kirksey (2003) reviewed the Civil Engineering Research Foundation's (CERF's) program. WEF is newer than AwwaRF and has not completed as many studies, although its work on sewer collection systems has clear parallels with AwwaRF work on distribution systems. In particular, technologies for condition assessment, planning and prioritization, and even renewal have similarities. CERF focuses on the construction side of utility work. M. Royer, a participant at two of the project workshops, explained USEPA research related to distribution systems to include distribution system modeling software (EPANET), disinfectant decay modeling, biofilm in distribution systems, costs, system failure case studies, leak and break detection and repair, decision support tools, corrosion and corrosion control, water quality effects in the distribution system (AWWSC 2002), intelligent systems conveyance and storage infrastructure, monitoring distribution systems via telemetry, and distribution system simulators. USEPA plans to make accessible a database of ongoing drinking water research projects. One source of information on USEPA's plans is the document titled *Research Plan for Microbial Pathogens and Disinfection By-Products in Drinking Water* (USEPA 1997).

CHAPTER 2
DISTRIBUTION SYSTEM ISSUES AND STATISTICS

CHAPTER OVERVIEW AND PURPOSE

The purpose of this chapter is to review the chronology of surveys about and case studies on distribution systems, statistics from distribution systems, and international comparisons. The chapter gives an overview of how research on distribution systems has progressed and how perceptions of issues have grown.

CHRONOLOGY OF SURVEYS AND CASE STUDIES OF DISTRIBUTION SYSTEMS

This section reviews major surveys about and case studies on distribution systems from the past two decades. Its purpose is to trace how knowledge and concern about distribution systems have evolved.

Before about 1960, a number of short papers were published about problems with CIP water mains. In those years, support for research was low and fewer scientific papers were published, compared to today, but utilities shared practical knowledge with each other. A paper by R.E. Morris (1967) in *Journal AWWA* is a good starting point for discussion of research on water main integrity. Morris's paper has been cited a number of times as an early exposition of causes and remedies of CIP breaks. As discussed later (in chapter 5), the paper illustrates that the author and his contemporaries had a good understanding of the role of factors such as corrosion, temperature, and construction practices on the performance of CIP water mains.

After the 1974 SDWA, in 1980, a U.S. General Accounting Office (USGAO) report raised awareness of the seriousness of distribution system issues but stopped short of recommending federal subsidies (Comptroller General 1980). The report illustrated that a few years after the SDWA was signed into law, distribution systems were being studied as a policy issue, just as the "infrastructure crisis" was unfolding. This crisis turned out to be not a crisis at all, but a media-driven yellow flag of caution that maintaining and renewing infrastructure was an expensive proposition that ought not to be ignored.

To document needs, more data on distribution systems were needed. In 1983, AWWA surveyed 39 of the largest utilities, later expanding the survey to create the Water Utility Database or AWWA's Water:\Stats product, which is available today (AWWA 1997, 2002). In reporting the early surveys at the 1985 AWWA conference, J. Gilbert stressed the need for capital management programs to sustain infrastructure (O'Day et al. 1986). However, the attitude among 33 of the surveyed large utilities was predominately that O&M expenditures were adequate to maintain reliability; although 23 of these utilities had reported some deferred maintenance. Apparently, concern about high levels of deferred maintenance and other concerns about distribution systems increased after 1985, especially after the SDWA Amendments of 1986.

A survey by CH2M HILL was reported at AWWA's 1985 Distribution System Symposium (O'Day et al. 1986). It showed that actions differ among small, medium, and large utilities. Percentages of reported actions are included in Table 2.1.

These statistics explain that large and sophisticated utilities have access to management expertise and programs that small utilities typically lack. However, reports like this do not show the extent to which management programs and technologies are actually used. This information

Table 2.1
Reported actions by utilities

	Small <5 mgd	Medium 5–50 mgd	Large >50 mgd
Leak detection surveys	16	19	40
Method to determine pipeline replacement need	63	77	100
Computer model of system hydraulics	48	58	90
Reports of main breaks	27	66	80
Steps to remove scale and tuberculation	13	31	30
Revenue to finance repair, rehabilitation, and replacement program	59	69	80
Budget and planning process for replacement	48	67	90

Source: O'Day et al. 1986

must be gleaned by in-depth case studies. All the surveyed utilities were increasing the use of management methods, and their actions depended on the availability of useful information to the utility. Generally, the surveys done in the 1980s showed that awareness of distribution system problems was increasing, and that utility practices varied significantly. This was also an era of new materials, products, and technologies, especially computer-related technologies.

The 1986 report by O'Day et al. (1986) has often been cited as an early comprehensive account of condition factors in distribution systems. The team surveyed a number of utilities, including New York, N.Y., Denver, Colo., Philadelphia, Pa., Louisville, Ky., East Bay, Calif., and Kenosha, Wis. The report contains information about failure mechanics, condition assessment, management methods, and other topics that are discussed later in this report.

Deb et al. (1990) studied Calgary, Alta.; EBMUD; the Los Angeles Department of Water and Power (LADWP); the Louisville Water Company; the Milwaukee Water Works; San Jose Water Company; and WSSC. In all, these authors surveyed 35 utilities about their renewal practices. Of these, 29 reported that they had a renewal program, and 5 of the 6 without a program were in the small-to-medium size category. Cleaning and lining were used by 70% of large utilities, and 38% of large utilities used more than one method but small ones did not. Nine of 11 small utilities used dig and replace, rather than rehabilitation methods. Most appurtenances were replaced rather than rehabilitated. Valves and hydrants were both replaced and rehabilitated. In planning, only 29% used a rating system to select mains for rehabilitation. Most water utilities kept records on renewal projects, with nine information items. From this report we see that repair, rehabilitation, and replacement were being practiced, especially in large utilities, but other than cleaning and lining, newer technologies were not in widespread use.

Kirmeyer, Richards, and Smith (1994) conducted a comprehensive survey of 20 utilities in 1992. They also summarized AwwaRF's three expert workshops on distribution systems, held in 1990, 1991, and 1992. This survey, and subsequent site visits, became the basis for a comprehensive research needs report, which is summarized in chapter 9. Issues raised in this report are also listed later in this chapter.

Deb et al. (2000) surveyed 37 utilities about O&M practices, including European utilities. Leak detection was the least common maintenance activity among tasks such as hydrant flushing and testing. Roughly 89% of the utilities had capital improvement programs. Of the utilities, 81% had corrosion control procedures in place and 60% had developed emergency procedures for main breaks. The most common information systems were infrastructure inventories and network models. About 70% had compiled maintenance history databases. Although the statistics show that the utilities engage in these activities, they do not reveal the extent to which they actually implement them. The survey also showed that information tools in the field and water quality modeling for distribution systems were not widely used by North American utilities.

Ellison's (2001) AwwaRF report about distribution systems synthesizes the literature up to about 2000. This report presents guidelines and answers to a range of questions about distribution systems. It is an excellent overview of the major issues and will be cited a number of times in this report.

While these surveys and case studies were proceeding, a number of needs surveys were also being conducted (see chapter 1). For example, AWWA surveyed 20 utilities of varying locations and sizes and reported needs of more than $250 billion over the next 30 years to replace pipes and infrastructure (*ENR* 2001).

In this project, we surveyed and collected information from approximately 50 utilities in the course of three workshops, several site visits, and additional mail surveys. Our results are consistent with previous studies, showing that technology use is advancing, but a gap remains between utilities that use sophisticated technologies and practices and others that do not. Examples of leading technologies and practices include

- Comprehensive condition assessment programs
- Geographic information system (GIS) software for system management
- Maintenance scheduling program and procedures
- Database systems to manage pipe component data and to track failures
- Capital improvement planning for water main renewal
- Use of trenchless technology for pipe rehabilitation

In summary, a basic engineering understanding of the *behavior* of distribution systems existed before the SDWA, but after the act and the onset of the so-called infrastructure crisis, much more attention was given to the *condition* of distribution systems. Technologies and options have increased rapidly, and management tools for distribution systems will continue to change.

STATISTICS OF U.S. DISTRIBUTION SYSTEMS

Statistics of U.S. distribution systems have been reported in a number of publications. AWWA's Water:\Stats database is the best overall source, but over the years various surveys have been conducted. USEPA has also issued some statistics. The statistics published by Kirmeyer, Richards, and Smith (1994) were more comprehensive than most, and are given below. Deb et al. (1998) also reviewed distribution system statistics, and their analysis was based on the 1991 Water Industry Database. As would be expected, the 1998 results did not differ much from those of Kirmeyer, Richards, and Smith (1994).

In sheer volume, the market for new pipelines and renovation of aging pipelines exceeded $20 billion in the United States in 2000, including potable water, sewer, stormwater, gas, crude

Table 2.2
Criteria for pipe replacement

Criteria	Percent reporting
Number of leaks or breaks	75
Age of pipe	45
Low flow	40
Condition or type or material	30
Size changes required	30
Water quality	15
Soil condition	15
Location	10
Street construction work	10
Elimination of dead ends	5
Amount of damage by leaks or breaks	5

Source: Kirmeyer, Richards, and Smith 1994

oil, and industrial pipelines (Jeyapalan 2000). Of this market, water supply occupies a major share. Kirmeyer, Richards, and Smith (1994) estimated that $2.8 and $1.7 billion were needed in 1992, respectively, for new and replacement pipe in the United States. New pipe per year was 13,200 mi, costing $2.8 billion. Types of pipe included DIP (47.7%), PVC (38.7%), and concrete pressure pipe (CPP; 12.5%). Scaling those numbers up by population increase—from 226 million in 1992 to 285 million today—would yield a U.S. water supply pipe investment (new and renewal) of about $6 billion per year. That suggests that of the overall $20-billion market, water supply constitutes about 30%.

More recent data are available on a limited basis, but the Kirmeyer, Richards, and Smith (1994) report still offers useful statistics. The authors estimated that 36.2% of O&M budgets, or about $4.5 billion per year (in 1992), went toward distribution and transmission. The top five expenditures were pipeline maintenance, appurtenance maintenance, pumping and transmission, storage, and water quality monitoring. Scaling this up by population and considering further aging would suggest a 2003 budget of about $6 billion for water distribution and transmission O&M.

In 1992, replacement was 4,400 mi of pipe each year, at a cost of $1.7 billion, plus disruption to customers and other utilities. Replacement rates varied from 0%–1.5% per year, with the average at about 0.5% per year (Kirmeyer, Richards, and Smith 1994). At that rate, it will take 200 years to replace the current inventory of pipe. How the age of inventory builds up is tracked using a Nessie curve, described in chapter 6. Table 2.2 lists criteria for pipe replacement.

Table 2.3 summarizes general statistics of U.S. distribution systems as of 1992 (Kirmeyer, Richards, and Smith 1994). Later reports corroborate the orders-of-magnitude of the statistics.

O'Day et al. (1986) presented a discussion of break rates, based on a survey of six cities. Data were inconsistent, and standard reporting formats were needed. Kirmeyer, Richards, and Smith (1994) estimated about 237,600 breaks per year in the United States, or 0.27 breaks/mi/yr.

Table 2.3
Statistics of U.S. distribution systems as of 1992

Estimated length of distribution piping	880,000 mi
Estimated replacement value of piping	$348 billion
Condition of piping	28% excellent, 43% good, 26% fair, 3% poor
Estimated number of pipe breaks	237,600 breaks/yr
Type of existing piping	48% CIP, 19.2% DIP, 15.1% A–C
Estimated new piping	13,200 mi/yr (47.7% DIP; 38.7% PVC; 12.5% CPP; cost $2.8 billion/yr)
Estimated replacement	4,400 mi/yr
Value of replacement	$1.742 billion/yr
Lead service lines	2.3–5.1 million
Cost to replace lead service lines	$10–$14 billion
Fire hydrants	5.85 million
% of O&M budgets to T&D	36.2
O&M budget to T&D	$4.5 billion/yr
Inadvertent system losses	10%
Cost of water losses	$2.8 billion/yr

Source: Kirmeyer, Richards, and Smith 1994

Deb et al. (2002a) also reported break rates, based on a later survey. They found from AWWA's Water:\Stats database that breaks in 1995 were 23 breaks/100 mi/yr. Also, break rates in Europe are higher, on the order of 50 breaks/mi/yr. In this project, we plotted break rates from the AWWA Water:\Stats data based on the 1997 survey, and the data show break rates in the same range, with great scatter in the data, especially for small utilities. Table 2.4 lists causes for main breaks.

On lead service lines, Kirmeyer, Richards, and Smith (1994) estimated that some 2.3–5.1 million require replacement, and to remove the utility portion would cost $3.4–$5.1 billion. To replace all would cost $10–$14.1 billion. Although lead service lines remain an issue, discussions in this project's workshops did not focus on it. However, a number of issues related to building plumbing systems will require attention in the future. (Note: AWWA's 2002 report on distribution system survey results has more recent data on service lines.)

INTERNATIONAL COMPARISONS

In the course of the project, we were able to learn about practices in a number of different countries. Of course, so many issues are involved that it is only possible to cite a few highlights.

In the Netherlands, AwwaRF has a partnership with KIWA, a Dutch water research association. KIWA has produced a Water Mains Systems Management protocol, which is being used to catalog and manage renewal programs. We did not gather detailed information on this program,

Table 2.4
Causes of main breaks

Cause	Percent reporting
Materials/deterioration	55
Weak joints	35
Earth movement or settling	30
Freezing	30
Internal corrosion	25
Corrosive soils	25
Construction or utility digging	25
Stray direct current	20
Seasonal changes in water temperature	15
Heavy traffic load	10
Tidal influences	5
Changes in system pressure	5
Water hammer	5
Air entrapment	5

Source: Kirmeyer, Richards, and Smith 1994

but it apparently arose from a research project intended to guide utilities in renewal programs and to offer general guidance that is similar to that used by major U.S. utilities (KIWA 2001).

In a report on benchmarking the Dutch water supply industry (VEWIN 2001), we learned that companies are beginning with asset management, but financial performance and customer satisfaction does not appear to have reached the level of distribution system integrity. The report states that

> A number of companies have started with the development of a structural approach for asset management. This means identifying, defining, and composing work activities for building, maintenance and work execution of existing and future production facilities and distribution networks, with the aim of minimizing the total costs across the entire lifetime of this infrastructure.

The CARE-W tool kit, a project of the European Commission, includes performance indicators, failure forecasting, reliability forecasting, and rehabilitation planning (Hafskjold 2003). Seven European countries participated in this project, including Germany, France, Italy, Portugal, Norway, the Czech Republic, and England. The research issues examined were condition of a specific pipe and the entire network, most vulnerable pipes, future failure rates, ranking of rehabilitation projects, future investments needed, and optimal management of water distribution networks.

In Scandinavia, a group developed a "Six Cities Report" on performance indicators (Stahre, Adamsson, and Eriksson 2000). The report specifies that performance indicators must be precisely defined, simple to measure, easy to control, easy to understand (by professionals and nonprofessionals alike), and as few as possible. The report proposes a set of performance indicators for the water industry comprising customer satisfaction, quality and availability, environment, staff and organization, and economy. It does not reach to the level of distribution system integrity, but a project reported at our project workshop did feature a review of the CARE-W management system for water distribution systems (described in chapter 6). One of the participating utilities in the study is the water utility in Sydney, Australia. Its report (Radovanovic 2003) is also described in chapter 6. Sydney has a hierarchy of capital plans, and the utility uses advanced models to carry out an impressive infrastructure management program.

Japan's experience was represented in the project in Matsushita's (2003) report, which focused on Kobe's experience with earthquake damage and recovery. Japan also has an active partnership with EBMUD and AwwaRF to share experiences about seismic improvement programs.

Canadian utilities face similar issues to those faced in the United States. Canada's Institute for Research in Construction and its Urban Infrastructure Rehabilitation Program are discussed in chapters 5 and 6.

DISTRIBUTION SYSTEM ISSUES

In the reports reviewed, a number of issues were raised. This section reviews them, and recommendations for further research are given in chapter 9.

To classify the issues, we turn first to the 1994 report by Kirmeyer, Richards, and Smith, where the investigators recommended the following changes in AwwaRF's issue categories related to distribution systems.

Infrastructure

The investigators recommended that this category of research be called "Water Distribution System Infrastructure." They thought extensive work was needed on testing, evaluation, and methods for renewal work.

Prevention of Water Quality Deterioration

Kirmeyer, Richards, and Smith (1994) recommended that the name change to "Water Quality Maintenance and Regulatory Issues." They also suggested a focus on regulatory changes.

Corrosion Measurement and Control

These authors recommended that this category be called "Corrosion, Permeation, and Materials Deterioration." They thought that the Lead and Copper Rule (LCR) would shift the emphasis and that a change from lab to field research was needed.

Distribution System Modeling

The investigators recommended that this be changed to "Distribution System Design and Operation" and that location and assessment system components should be stressed.

These area titles work well for classifying issues and are used in the discussion that follows. However, the focus of this report on distribution system integrity requires that the topic of "infrastructure" be expanded, so the following categories are adopted for the discussion of issues:

- Infrastructure management
- Planning, prioritization, and performance evaluation
- Pipe materials and structural research
- Condition assessment and failure mechanics
- Distribution system engineering, design and operation
- Maintenance management
- Construction and renewal
- Corrosion measurement and control
- Water quality maintenance and regulatory issues

Infrastructure Management

To pick up strategic issues, we added the topic of infrastructure management. We did not adopt the combined term "corrosion, permeation, and materials deterioration," preferring the term "failure mechanics." Corrosion is a very specialized field, and failure mechanics needs to be identified separately, so as to collect recommendations about studying it and organizing data in a database. Major issues we identified are described briefly here.

High Investments Required. Needs studies project large sums required to repair, rehabilitate, and replace distribution systems. However, their credibility rests on the standards they reflect, estimates of condition, and service lives, all of which are uncertain. Although evidence for the need to improve distribution system infrastructure has increased, it requires more rigorous analysis. The lack of rigor in proving needs is based on the issue noted by O'Day et al. (1986): lack of standards and measurement procedures.

Workforce. Workforce has not been highlighted as an important issue, but our project identified associated issues to be discussed in chapter 8.

Planning, Prioritization, and Performance Evaluation

Since the level of concern about distribution systems began to rise in about 1980, improved planning methods for renewal have been highlighted as an issue. One important concern is uncertainty about the expected service lives of water mains. Without knowing how long water mains last, it is impossible to accurately estimate investment needs.

Pipe Materials and Structural Research

The need for research about pipe materials and structural characteristics has been cited on a continuing basis in workshops and during projects.

Condition Assessment and Failure Mechanics

Condition assessment is an area of great promise for distribution systems management. The knowledge base for water main failure mechanics needs further study. High-quality water is required at the tap, but there is a shift in responsibility at the property line and service connections, and plumbing systems are handled by a different industry than public water supply systems. AwwaRF reissued a request for proposal (RFP) in 2003 for research about service connections and fittings.

Distribution System Engineering, Design and Operation

The Kirmeyer, Richards, and Smith (1994) report recommended emphasis on tools for locating and assessing condition of system components, as well as on performance standards, control systems, models, and mapping. We recommend condition assessment as a separate category, but we found that a number of engineering issues need attention as well.

Maintenance Management

Maintenance management continues to be a strategic issue for distribution systems, although utilities have made considerable progress in the last decade, particularly with computerized maintenance management systems (CMMSs).

Construction and Renewal

We combined the topics of construction and renewal because they involve contractors, the construction process, machinery, and construction industry advances. Figure 2.1 shows a typical utility construction scene. Much progress has been made in this area, as discussed in chapters 7 and 8.

Corrosion Measurement and Control

The specialized issue of corrosion is costly, with annual costs of about $275 billion per year (1998 basis) and costs to utilities of about $48 billion per year (Bianchetti 2001) in the United States. AwwaRF has funded a number of proposals in this area, as summarized in chapter 5.

Water Quality Maintenance and Regulatory Issues

Although water quality continues to be a main area of concern for distribution systems, it was not a focus area in this project. However, we briefly review the issue in chapter 5.

Research Policy

Chapter 9 summarizes some research issues for distribution systems that go beyond the realm of research topics and extend into policy.

Figure 2.1 Water main construction

CHAPTER 3
WATER MAIN TECHNOLOGIES

CHAPTER OVERVIEW AND PURPOSE

Of the technologies involved in distribution systems, the manufacture of water mains themselves influences their integrity the most. This chapter presents a chronological record of innovations in pipe materials and the increase in the knowledge base about water mains. Other chapters deal with how pipelines are designed and built, how condition is assessed, and how failures are diagnosed. This chapter does not go into detail about manufacturing processes, standards, failure mechanisms, and related knowledge, but it supplies references to this information. Topics included in this chapter include

- Early history of pipe technology
- Modern piping
- Water utility experiences with pipe materials
- Choices for distribution system piping

EARLY HISTORY OF PIPE TECHNOLOGY

Pipe technology dates back several millennia. Mays (2002) explains how water was brought into homes in the Indus and Mesopotamian civilizations in the 3rd millennium B.C. and displays a photograph of a distribution pipe surviving from about 2000 B.C. Early systems included lines of pipes made of stone and lead. Of course, the Roman aqueducts are famous, but other civilizations built impressive systems, too (Frontinus 1973 [translation]).

Although water distribution technologies in early civilizations date back thousands of years, the European systems that preceded current U.S. practice came later. It is natural to expect that early technologies from the Mediterranean area were imported to Europe, and water and wastewater systems of the 13th century A.D. have been noted in Swedish towns (Drangert, Nelson, and Nilsson 2002). These early settlements lacked sophisticated organization, and water supply was a private responsibility with no public funding. Organized water systems were usually only for the fortunate few; the masses lived without them.

The earliest known CIP, in Versailles, France, dates from 1664. Reports vary about the earliest U.S. systems, but Mays (2000) reported that the first piped supply was constructed in Boston, and the first piped community supply was built in 1746 in Schaefferstown, Pa. An early water supply system in Bethlehem, Pa., dated from 1754. It used spring water forced by a pump through bored logs. Philadelphia also developed a water supply system, with combined public and private pumping facilities. Initially, the pumps were usually horse-driven and were later driven by steam before electric power was available. During the 19th century, many communities obtained their water from nearby springs and wells that were easily contaminated. New York's water supply story begins in about 1800 with the Manhattan Water Company, which opened the Croton Aqueduct in 1842. In the 19th century, steam power and pumping made possible the delivery of water under pressure.

Before 1900, U.S. cities began to install underground pipes for water distribution. Mays (2000) dates the first CIP to 1817 in Philadelphia. U.S. cities such as New York and Washington

Photo of historical Korean picture by N. Grigg

Figure 3.1 Early water main construction, Seoul, Korea, in 1908

had significant lengths of buried CIP in service by 1900. Clift (2001) has written a book about pipe history. Figure 3.1 shows comparable turn-of-the-century pipe construction in Seoul, Korea.

MODERN PIPING

As a result of the industrial awakening, new technologies began appearing during the 19th century. The invention of electric power systems and gasoline engines opened new possibilities for distribution systems. By 1900 the United States was urbanizing rapidly, with water systems mostly constructed of CIP, but with some made of riveted steel, wood staves, and wrought steel. AWWA has published a number of texts and standards about water main piping. The most recent texts on water transmission and distribution give information on pipe types (AWWA 2003).

In 1900, CIP was "pit cast," meaning that it was cast in vertical sand molds. Quality control was a problem, but pipe walls were thick and the pipes did not break easily. In 1908, AWWA published the first standards for vertical pit casting of CIP, and this represents the first recognizable standard in AWWA's files (F. Baruth, personal communication, 2003). In 1921, centrifugal casting was introduced. Metallurgy was improving, and tensile strength became higher. However, wall thickness was reduced.

In 1948 DIP was developed by adding an inoculant, usually magnesium, to cast iron. This enables the graphite to distribute itself differently in the metal, making the pipe stronger and tougher than CIP. Usually, a thin lining of cement is added to DIP to increase its resistance to internal corrosion. A coating of bitumen and a PE wrap is added to resist external corrosion. National standards for DIP were introduced in 1965. The Cast Iron Pipe Research Association later became the Ductile Iron Pipe Research Association. A survey done by the association in the late 1960s showed that in the United States about 90% of the larger mains, 4 in. and larger, were cast iron.

Figure 3.2 Concrete pipe

Until the 1920s, the most common joint was of lead in a bell and spigot for CIP. From 1920 to about 1955, some cities used a sulfur compound to replace lead, but it was too rigid and led to leaks. Later, a mechanical joint was introduced, then a rubber ring gasket joint, where the gasket fits in a groove in the bell end of the pipe.

Concrete pipe was used for sewerage in the Roman system. The earliest use of concrete pipe in the United States was in New York in 1842, for sewer pipe. Reinforced pipe was developed in 1905, prestressed pipe in the 1930s, and steel-cylinder prestressed pipe in 1942. Use of concrete pipe has been growing since 1950. It is available as prestressed concrete cylinder pipe (PCCP), pretensioned concrete cylinder pipe, reinforced concrete cylinder pipe, and reinforced concrete noncylinder pipe (see Figure 3.2).

Steel water pipe has been used in the United States since 1852. Its interior is usually protected by either cement mortar or epoxy (see Figure 3.3). Today, it is mainly used for larger lines. Cates (1971) wrote a history of steel pipe.

A–C products, sometimes called cement fiber, were probably developed in Russia around 1896. A–C pipe was developed in the 1920s and introduced in the United States around 1930. Although studies have not identified a hazard in its use for water conveyance, asbestos is being phased out in the United States because of hazards from airborne fibers (Australian Academy of Technological Sciences and Engineering 2003).

PVC was first developed in 1835 in Germany, and the oldest known PVC pipe was produced during the 1930s, also in Germany. Plastic pipe was first introduced in the United States around 1940. PVC pipe is the most widely used plastic pipe for smaller water mains. The National Sanitation Foundation started to certify it in the 1950s, and in the 1960s and 1970s, the American Society for Testing and Materials and AWWA developed standards. Today, a large share of small-diameter pipe is PVC.

Courtesy of City of Sioux Falls, S.D., Public Works Department

Figure 3.3 Sample of cement-lined pipe

PE pipe emerged more recently from the synthetic plastics industry as a water pipe, and AWWA has published standards for it as well. Although it is a new product, utilities are showing interest in using it. Use of PE and polybutylene (PB) pipe has mostly been for small-size applications, but interest has been growing in using the larger sizes.

WATER UTILITY EXPERIENCES WITH PIPE MATERIALS

Much of the literature about CIP is about failure mechanics and is discussed in chapter 5. Our workshops also produced experiences with concrete pipe. Experiences with A–C pipe were reported to be generally good, but actions associated with it are generally to replace it, given the policies followed by most utilities.

Recent studies have largely focused on evaluating plastic pipes, and AwwaRF has funded several studies on plastic pipe performance. Thompson and Jenkins (1987) wrote *Review of Water Industry Plastic Pipe Practices,* with a goal of reviewing water industry practices for plastic pipe and summarizing information on permeation. The report contains statistics on the PE, PB, and PVC piping of that time, along with a customer survey. It describes problems with PE and PB, such as brittle failure, installation, fittings, and bends. These investigators concluded that metal services were more satisfactory than plastic at that time, but today, utility representatives are more satisfied with PVC mains (permeation can be predicted through PE and PVC pipe but not PB pipe). AwwaRF is currently studying the permeation problem.

The goal of the 1994 report by Moser and Kellogg was to evaluate the use and performance of PVC pipe. A survey was done where each utility supplied a sample of PVC pipe to be tested. The conclusions discuss reasons for failure and note that almost 50% of problems reported with PVC pipe occur in the first year after installation.

In our project workshops, utilities personnel reported that experiences with PVC have improved since the Thompson and Jenkins report in 1987.

Thompson and Weddle (1992) wrote about plastic service lines. Their goal was to provide a link between the reported causes of dissatisfaction with PB and PE pipe and specific pipe and

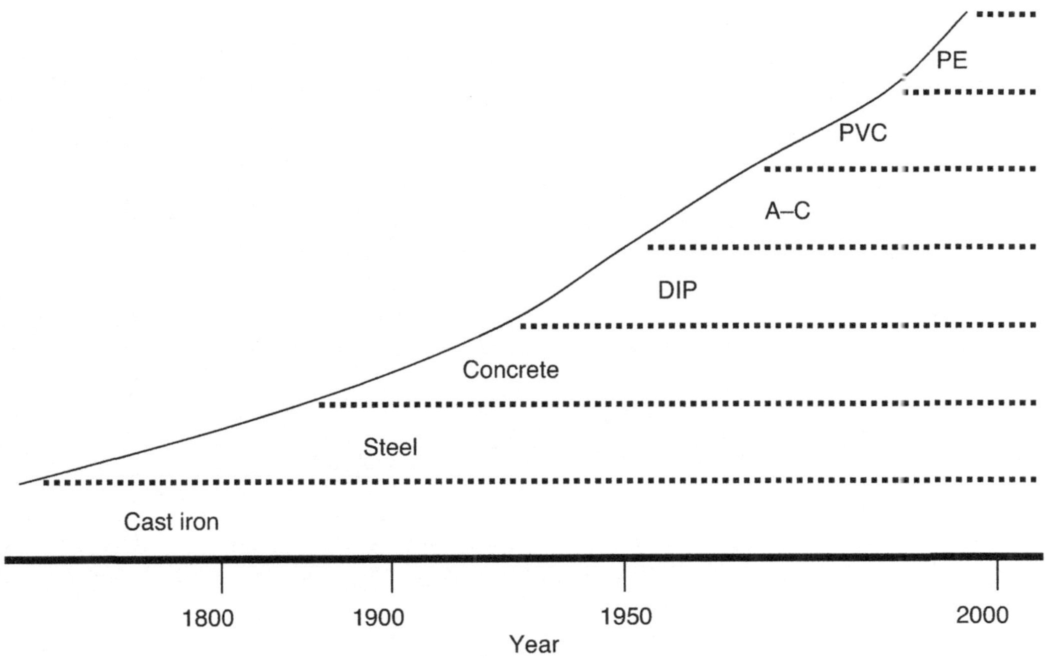

NOTE: The diagram is conceptual only. Years are not precise and not all materials are shown.

Figure 3.4 Development of pipe materials

fitting materials and installation practices. Plastic service lines have resistance to corrosion, electrical nonconductivity, relatively low cost for materials, and ease of installation. For the study, the investigators conducted interviews and carried out a laboratory program using the samples of failed pipe submitted by utilities. Thompson and Weddle found that the predominant causes of failure were brittleness, bedding, installation, and kinking.

CHOICES FOR DISTRIBUTION SYSTEM PIPING

The number of choices of pipe materials increased steadily during the 20th century. Before about 1950, almost all distribution pipe installed was a form of cast iron, but today utilities can choose other types of pipe for different kinds of installations, including new construction and renewal. Figure 3.4 presents a conceptual overview for the increase in pipe materials available.

Smith and colleagues (2000) devote an appendix to a "piping compendium," which describes characteristics for a list of pipe types. Characteristics include material, corrosion resistance, cost, and many other design parameters. In a presentation at the Annual Conference and Exposition (ACE) 2003, R. Hope included a graph of pipe inventory showing the lengths of pipe installed during each decade according to material and size. He also presented a graph of survival functions for each type, demonstrating how long each type was expected to last. He thought that we probably only know about 60%–70% of what is in the ground. In his second presentation at ACE 2003, Hope displayed another chart that showed the distribution of pipe material according to geographic region in the United States and Canada.

CHAPTER 4
PIPELINE ENGINEERING AND CONSTRUCTION

CHAPTER OVERVIEW AND PURPOSE

This chapter presents the engineering and construction technologies that determine how well pipelines are designed and built. Although the technology of pipeline engineering is a practical field, without a large base of research projects and scholarly publications, it may hold keys to improved distribution system integrity. Reviewing knowledge about engineering and construction in the pipeline field is more difficult than in more focused fields, but our review may reveal areas where research or changes in practice could enhance integrity of pipelines. The main point is that research into design and construction processes might yield unanticipated benefits. This chapter covers the following areas:

- Pipeline engineering overview
- Pipeline engineering evolution
- Pipeline design
- Construction techniques
- Pipelines in the oil and gas industry

PIPELINE ENGINEERING OVERVIEW

Pipeline engineering is a niche field shared by several disciplines. Selig and Nash (1988) write:

> Buried pipe infrastructure by its very nature is not a highly visible field of engineering, and not one of major public awareness or glamour. The participants are often relatively small companies or government agencies. For these reasons the extent of technology growth as represented by research activity has been relatively small. An important conclusion from the workshop is that a greatly increased research effort is both needed and justified by economic considerations.

About pipeline engineering, Moser (2001) writes, "City managers and pipeline engineers are sobered by the present-day reality of deteriorating pipe systems. The problem is almost overwhelming." Moser attributes the beginning of pipe system theory to Marston's landmark paper (Marston 1913) and Spangler's paper on flexible pipe (Spangler 1941). Moser's section on pipe installation and trenchless technology includes the same topics as AWWA's Manual M28 (AWWA 2001): transportation, trenching, installation, thrust blocks, bedding and backfill, and newer methods based on trenchless technologies.

Even though water distribution systems are unique, some technologies—especially those for condition assessment and repair, rehabilitation, and replacement—can be transferred from other fields. For example, oil and gas pipelines operate under pressure, like water pipes, and they also require inspection and repairs. Sewers do not normally operate under pressure, but they involve similar design, workforce, and construction issues, so some technologies may be transferable. Even planning and other "soft" technologies can be transferred from public sector fields, such as public works, which use facilities software that can be adapted to water utilities.

Courtesy of Sioux Falls, S.D., Public Works Department

Figure 4.1 Pipeline construction

Sectors where pipelines are most important include water supply transmission and distribution, municipal wastewater, oil transmission, and gas transmission and distribution. Other pipeline applications include carbon dioxide transport; drainage, culverts and subdrainage, and storm sewers; hydroelectric penstocks; industrial wastewater; irrigation; pneumatic tube conveyance; power and communication conduits and utility tunnels; power plant fluid systems and cooling water; steam distribution; and materials transport (capsule, slurry, stripping; Selig and Nash 1988). Figure 4.1 shows a typical pipeline under construction in Sioux Falls, S.D.

From a professional viewpoint, pipeline engineering has branches in the disciplines of mechanical, civil, and chemical engineering. ASCE has a Pipeline Engineering Division, which sponsors periodic technical conferences that include advances in water pipeline practice. The American Society of Mechanical Engineers also sponsors conferences and publishes reports that emphasize process piping and oil and gas.

PIPELINE ENGINEERING EVOLUTION

From an overall vantage point, then, pipeline engineering can be considered a fragmented field, with some transfer of knowledge from one field to another, but mainly a field where techniques are mostly unique to applications, such as water supply, oil and gas, and so forth.

Given the status of pipeline engineering, we can characterize its evolution as shown in Figure 4.2. It begins with basic construction techniques, which date back many years, with higher productivity arriving along with construction machinery. Then, as knowledge increased, it was published in handbooks and textbooks, and then in codes and standards. The field of pipeline engineering emerged as knowledge increased, and then computers and nondestructive evaluation (NDE) techniques came along, adding greatly to the technology level of the field.

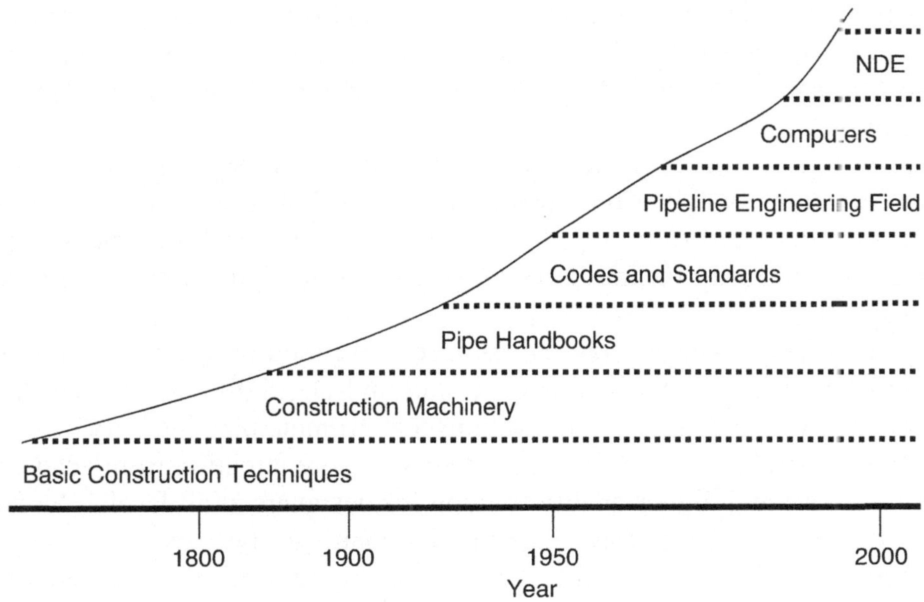

NOTE: The diagram is conceptual only. Years are not precise and not all developments are shown.

Figure 4.2 Evolution of pipeline engineering and construction

Perhaps because the field of pipeline engineering is fragmented and practical, it has few "knowledge generators" to unify it, such as research institutes, textbooks, academic departments, and conferences.

PIPELINE DESIGN

In this section we discuss the tasks of the field of pipeline engineering and construction to discern how technologies have evolved and where future payoffs might be. The first phase of pipeline engineering is the process of engineering design and its associated tasks of planning, design details, construction management, coordination with other utilities, and damage prevention. All these tasks are important to quality in construction, which improves the integrity of constructed pipelines.

Because the design process is important to the integrity of pipes, skilled designers ought to make decisions on critical elements. However, much pipe is designed by generalists and consists of little more than layout drawings, without special investigations. As AWWA (2001) Manual M28 states, "detailed plans of the project should be prepared by the project engineer. These plans should include the details of the alignment, grade, and depth of mains, the locations of valves, hydrants, and fittings, and details of all known obstructions." Later the manual points out that in some cases, the bedding ought to be prescribed by the design engineer. The point is, design may go further than just straightforward specification of location parameters and pipe types. Training for designers may be another key factor in prolonging pipe life, and research might evaluate how well designers are trained. However, as pipe failures may occur well into the future, the feedback between good design and proof of good design might be a long time in coming.

A range of approaches to pipe selection can be taken, but sale of pipe is a competitive process, and no definitive guideline can be published without being anticompetitive. Smith et al.

(2000) reports that DIP is currently favored by utilities and is installed only by open-cut methods. However, new joints are now being developed to allow the use of DIP in trenchless methods. Indications from our research are that utilities are considering a range of materials and methods with different types of pipe.

Pipe inspectors should know the critical factors for long-term integrity. In our project, we heard how investigations of failed pipe detected construction problems, such as dropped pipe, scratched pipe, pipe damaged from impacts of machinery, and so forth. Unless inspectors detect these problems, pipe integrity cannot be ensured. Research might evaluate the effectiveness of construction inspection.

Coordination with other utilities and damage-prevention programs also fall into the realm of pipeline engineering. Programs such as One-Call and Dig-Safe have proved their worth and should be mentioned when identifying innovations that promote pipe integrity.

Education and training for pipe design deserves more attention from academics, to result in more comprehensive and higher quality training for designers at all levels. Research might be carried out on construction effectiveness in the utility contractor business.

Once needs are identified and specific segments of the system are targeted for rehabilitation or replacement, it is important to gather as much information as possible about the options available to solve the problem. Pipe rehabilitation methods, which include a number of emerging techniques, are described in chapter 7. Designers ought to be aware of new methods and their advantages and disadvantages, and how well such new information is disseminated might be a subject for research.

Cost factors for different methods also require investigation. Selvakumar, Clark, and Sivaganesan (2002) summarized anticipated costs of different methods.

In pipeline design, a number of decisions are required: conventional or narrow trench; type of pipe and pipe characteristics; depth, bedding, and backfilling; installation methods; distance from other utilities; joint assembly; and cathodic protection or retrofit. The designer must be aware of choices, costs, impacts, and construction difficulties related to these decisions.

Reliance on new methods based on databases and computer-aided design (CAD) should enhance the capability of designers to specify materials and construction practices that can solve local problems. For example, designers should be able to gain experience with different environmental conditions in a city by linking their investigations with failure mechanics databases managed by other sections of a utility.

Construction management may hold secrets to improving pipeline installation. For example, in using cement–mortar lining (CML), utilities found that bundling projects and using experienced contractors worked to their advantage. Finally, there is the full realm of construction research, where research might improve contracting methods, construction techniques and methods for specific tasks such as excavation and trenching, bedding and pipe laying, and backfilling. Most exciting, however, is the possibility for new machinery and trenchless technologies.

CONSTRUCTION TECHNIQUES

Construction tasks involved in pipe installation also offer opportunities to improve pipe integrity. These encompass pipe selection, shipment, handling, excavation (see Figure 4.3), pipe laying, backfilling, and pressure and leak testing. All these construction tasks were mentioned during our project as affecting pipe integrity, problems, and failures. Rather than appearing in the

Courtesy of City of Carlsbad, N.M.

Figure 4.3 Open trenching project

research literature, construction techniques are usually discussed in textbooks such as *Water Transmission and Distribution* (AWWA 2003).

Pipe shipment and handling are important to pipe survival and avoiding failure. Experienced engineers report that investigating failures sometimes suggests problems with pipe caused by handling, rather than materials or actual construction. The work processes of shipment, inspection, unloading, stacking, and stringing can lead to pipe damage unless carried out properly. An issue to be studied is how well the workforce is prepared for these tasks. Training for inspectors and regulatory methods for pipe handling might be subjects for further study.

Excavation and trenching are not thought of as research subjects, but AWWA's (2001) manual states, "by far the most expensive part of pipe installation is the excavation. The trench should therefore be as shallow and as narrow as safety and soil conditions permit." These are, then, important subjects in pipeline integrity, and they will become more important as urban areas become more congested. The cost of excavation and trenching may become a driving force in the promotion of trenchless technology.

Technologies for digging and trenching are mature. In recent years, emphasis has been placed on the One-Call and Dig-Safe programs. A great deal of information about digging and location is available from the Electronic Exhibit Hall on Safe Digging (n.d.) and the National Utility Locating Contractors Association (2004) Web sites. The market for vacuum excavation has decreased recently because of the decline of the fiber cable market, which involved a great deal of horizontal directional drilling.

By the same token, bedding is an important subject that determines aspects of pipe integrity. Methods and specifications for pipe bedding might be reviewed, with model specifications for different circumstances produced. According to AWWA's Standard C600-99, *Installation of Ductile-Iron Water Mains and Their Appurtenances* (AWWA 1999), it covers the "minimum requirements for the installation of ductile-iron water mains and their appurtenances, including materials, dimensions, tolerances, and testing procedures. Installations that require special attention, techniques, and materials are not discussed." The standard was not reviewed as part of this project, but our work suggests that a review of guidance for pipe bedding might be in order.

Tunneling is another construction subject that appears in a review of pipe integrity. Microtunneling is, of course, a fast-advancing field of construction, with great promise for utilities and pipe installation and renewal. It was not reviewed under the project, but is listed here as a possible field for future research, and will be discussed later in the chapter on renewal methods.

Other design and construction topics that affect pipe integrity and might benefit from research attention include thrust restraints, backfill, and testing and inspection.

A paper by Clark et al. (2002) is a reference for the work required to install water mains. The authors list cost functions for the following tasks, all of which enter into the discussion of pipeline integrity:

- Base installed pipe—Pipe, gasket, joining material, and labor and equipment to lift the pipe into the trench and complete the joining operation
- Trenching and excavation—Labor and equipment costs for trenching and excavation
- Embedment—Placing materials under the pipe, between pipe and walls of trench box, and immediately above the pipe
- Backfill and compaction
- Valves, fittings, and hydrants—Materials and installation
- Sheeting and shoring—Labor and equipment costs
- Horizontal boring—Jacking of pipe underneath obstacles, includes casing and grouting
- Pavement removal and replacement—Cutting and replacement of pavement
- Utility interference—Overcoming interference from other utilities
- Traffic control—Labor and equipment
- Household service connections—Labor, equipment, and materials
- CML
- Sliplining
- Corrosion control

PIPELINES IN THE OIL AND GAS INDUSTRY

During the project, we reviewed research in the fields of oil and gas to determine if their technologies should be studied further for possible use by water utilities. Although our search was not exhaustive, our findings may be of some interest to readers of this report.

Oil and gas are, for the most part, private-sector activities, and research findings do not appear as widely in public or government-sponsored documents as those of the water industry. Consequently, there seem to be fewer technical manuals, journals, and textbooks, especially those that professionals in water utilities would use. One text that provides an overview is by Kennedy (1993).

On its Web page, the Gas Technology Institute (GTI; formerly the Gas Research Institute), describes itself as the "leading research, development and training organization serving energy and environmental markets" (GTI 2004). We found that a number of GTI's reports deal with gas pipe rehabilitation, especially re-lining methods.

In the federal government, the U.S. Department of Transportation's (USDOT) Office of Pipeline Safety (2002) conducts research programs on pipeline technology, operations, controls, and monitoring. The office's research relating to pipeline integrity has implications for the water industry and requires further investigation. Recent topics studied include

- Damage prevention and leak detection (internal and external monitoring devices to detect stress corrosion cracking—USDOT investigators believe that current inline inspection technology yields a high false-positive detection rate for stress corrosion cracking, and they propose studying occurrence and impact patterns, ultrasonic technologies for crack detection, and other technologies such as electromagnetic acoustic transducers, internal and external pipeline monitoring devices, and combined acoustic/electromagnetic sensors)
- External corrosion control (cathodic protection systems; monitoring of stray electrical systems resulting from coating anomalies; commercially available coatings under various conditions; improved coatings to impede environmental influences; seasonal variations and models for predicting corrosion wall loss)
- Internal corrosion control (corrosive products or microbes; induced corrosion control inhibitors; environmentally benign biocides to control corrosion induced by microbes; internal corrosion direct assessment methodologies; system operating conditions; corrosion-inducing bacteria identification)
- Risk assessment (data on the condition and integrity of pipelines; realistic maintenance priorities; computer programs to analyze risk assessment data; standard methods of risk and reliability analysis; and key factors such as welds, repair materials, surface loadings, material, and mechanical properties)
- Human factors (rotating shifts; human fatigue; interaction between operators and machinery; operator error minimization; human interfaces)

This office recently solicited white papers on improved materials performance to focus on improving pipeline materials to extend the integrity and lifetime of installed pipelines and their various components. Topics include damage- and defect-resistant materials, higher grade/strength steel, higher pipe design pressure, welding and joining, composite pipe, and pipe coatings.

In summary, oil and gas engineers have developed many pipeline technologies, some of which, if the economics are right, can be used in the water field.

CHAPTER 5
CONDITION ASSESSMENT OF WATER DISTRIBUTION SYSTEMS

CHAPTER OVERVIEW AND PURPOSE

This chapter describes methods for assessing the condition of water distribution systems and their components. Utilities need condition assessment information for engineering, O&M, and financial decision-making purposes. Although formal condition assessment is not practiced by all utilities, it holds promise for the future of distribution systems management. However, some utility managers believe that a reactive approach to renewal is best (i.e., emphasis on fixing systems without trying to anticipate problems through condition assessment). Not using condition assessment to anticipate problems is one extreme in a continuum that seeks to minimize breakdowns and optimize cost-effectiveness. Terms used are "breakdown maintenance," "on-condition maintenance," and "reliability-centered maintenance (RCM)." RCM is a proven concept in other industries, and AwwaRF is continuing to study the topic.

Research suggests that important advances will be made in condition assessment. The networks of aging pipes distributed over vast areas will require renewal with relatively low investments but high stakes. Therefore, condition assessment can be a powerful tool for deciding on courses of action.

This chapter begins with an explanation of a framework for the different ways of conducting and using condition assessment. Next, it discusses indicators that can be used to report condition. The framework is linked to failure modes because failures make the pipe unable to carry out its function, so failure mechanics is discussed next. Topics covered in the chapter are

- Condition assessment and its tasks
- Evolution of condition assessment
- Uses of condition assessment information
- Framework for condition assessment
- Failure mechanics
- Methods for pipe condition assessment (components)
- Methods for distribution systems assessment (systems)
- Data systems for reporting condition of components and systems
- Corrosion science and control
- Water quality and contaminant monitoring
- Project workshop examples and report

CONDITION ASSESSMENT AND ITS TASKS

What is condition assessment? The condition of something means a state of readiness or physical fitness for a given purpose. Condition of water mains determines how well components and systems can perform their functions. Assessment is the process of evaluating, so condition assessment means to evaluate the readiness of a component to perform its function. Ellison (2001) and Rajani (2003) give overviews of condition assessment.

Condition assessment requires analysis to break the problem into parts to examine them and diagnosis to identify the cause of any problem. Thus, in condition assessment, we study the

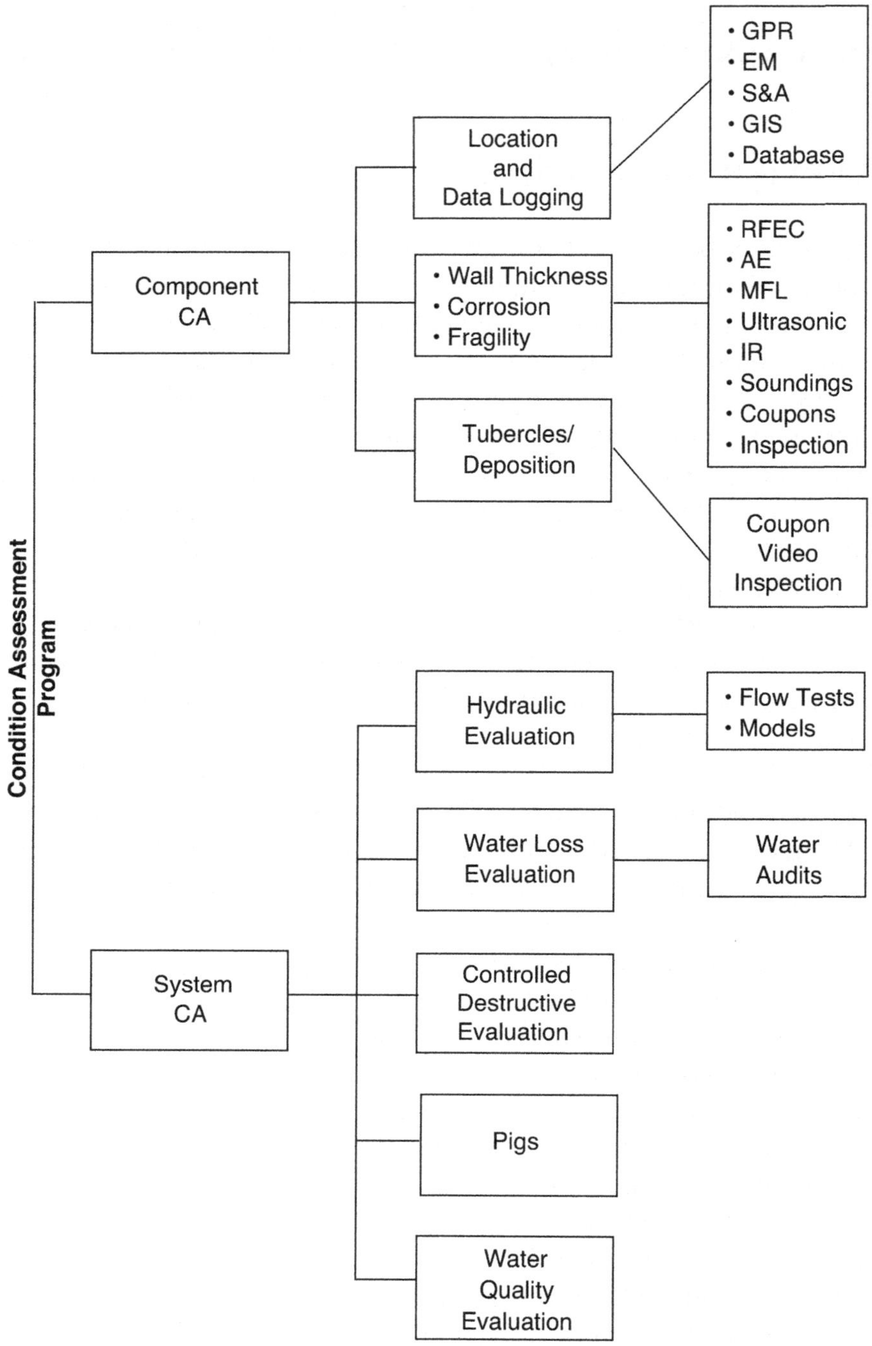

AE = Acoustic Emission
EM = Electromagnetic
GPR = Ground-Penetrating Radar
IR = Infrared
MFL = Magnetic Flux Leakage
RFEC = Remote Field Eddy Current
S&A = Sonic and Acoustic

Figure 5.1 Decision tree for condition assessment

parts and causes of problems and make assessments from the evidence. Condition assessment of pipe systems, then, includes ratings about condition and why condition might be good or bad.

Although many devices and scenarios can be used for condition assessment, the decision tree is fairly compressed because the technologies have similar functions—to assess some aspect of either hydraulic or structural condition. In Figure 5.1, which shows the decision tree, the tasks of condition assessment divide according to whether a component or a system is to be assessed. If a component is to be assessed, the basic tasks are location and inventory, wall thickness (structural strength), and tubercles/corrosion indicator. If a system or zone of a system is to be assessed, the basic assessments are hydraulic, water loss, and controlled destructive evaluation.

Each of these assessment categories has available for it several technologies, also shown on the decision tree.

EVOLUTION OF CONDITION ASSESSMENT

The roots of condition assessment for pipes and hydraulic systems are in the engineering and scientific methods that have evolved over time. During the 19th century, advances were made in fluid mechanics and hydraulics; today, we have sophisticated methods for network analysis and water auditing. In the field of structures, early understanding of materials and stresses led to better theories of pipe behavior and response. Chemistry has taught us about both internal and external corrosion, as well as about water quality. Early development of relatively simple measuring devices, such as flowmeters and pressure sensors, has given way to advanced instrumentation to measure many variables of water quantity and quality. New technologies using electronics, optics, radar, and acoustics have produced many possibilities for advanced instrumentation, and new fields such as corrosion science and forensics have emerged. Combining new technologies with computers leads to many possibilities, such as correlation devices. Computers also give us database technologies that can store and retrieve information rapidly.

The limit on advances in condition assessment is not technological but economic. In many ways, assessing the condition of an underground pipe network is not unlike assessing health conditions in a person; for example, examining the condition of coronary arteries or discovering the presence of a tumor. Humans are willing to spend large sums to diagnose a single incidence of illness, especially when insurers are paying their bills. For pipes, however, equivalent sums for diagnosis are not available, despite the fact that a segment of pipe might have a high replacement value. Considering that replacement value of water mains is high, they will increasingly be viewed as important assets worth spending money on for condition assessment. If a length of main would cost, say, $1 million to replace, what is a reasonable sum to spend on planning for its maximum usage? The risk of neglecting it or replacing it too soon certainly justifies reasonable investments in condition assessment. The lack of motivation to invest in condition assessment is caused by the stakes involved—the pipe is not going to "die," usually, and it may remain in service in spite of its deteriorating condition.

AwwaRF's review of condition assessment, *Noninvasive, Nondestructive Assessment of Underground Pipelines*, is optimistic about how NDE-based condition assessment can help in management (Dingus, Haven, and Austin 2002). The authors wrote: "a shift from empirical data to NDE methodologies will allow system operators to make optimal decisions based on pipe condition, leading to rehabilitation and replacement of damaged pipes only." The report finishes with an optimistic statement:

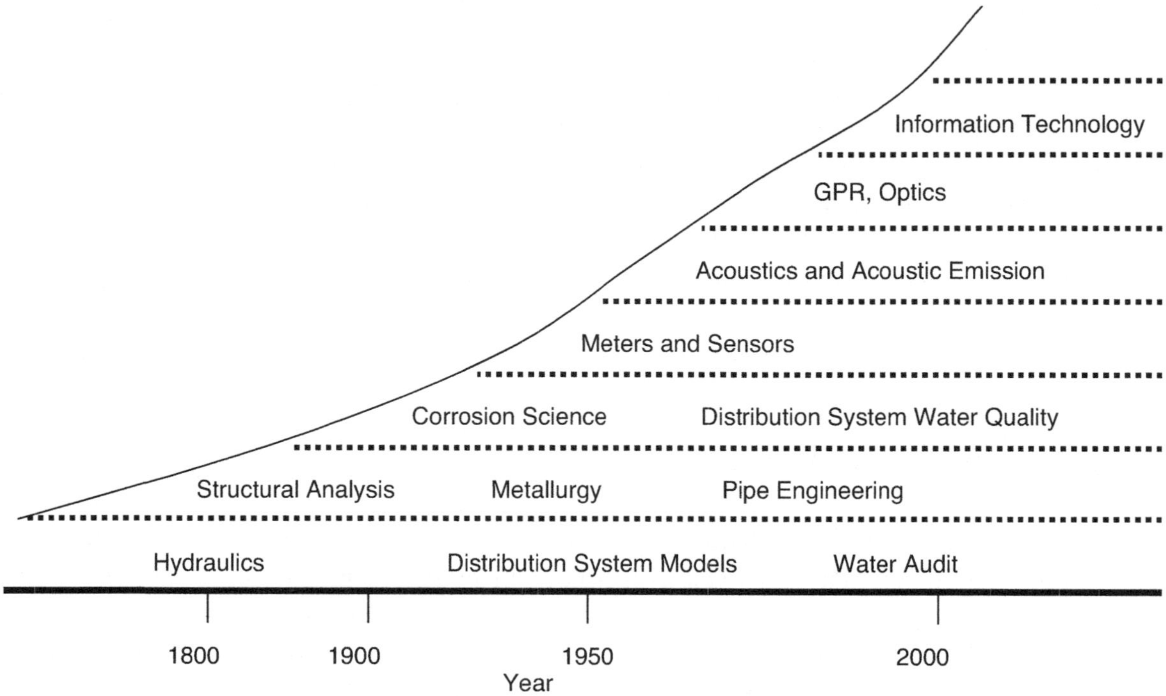

NOTE: The diagram is conceptual only. Not all technologies are shown.

Figure 5.2 Evolution of analysis and assessment methods

> NDE of underground pipes is paving the way toward a more efficient and cost-effective water utility. Not only will water utilities be able to find pipes that are already in a severe state of deterioration, but some methods, such as passive acoustic, also allow real-time data on the condition of pipes....Future NDE will allow utilities to examine pipes and create a life expectancy timetable from the data....No method for predicting failures will ever be 100% accurate, but this will give a closer approximation than...today's information.

The spectrum of possibilities for condition assessment is broader than NDE, of course, but these optimistic statements lead to a question for water utilities: Can utilities improve their capital management strategies with more active NDE techniques, or should they continue to make decisions based on multiple sources of field-based empirical information?

Taken together, the technologies available for pipe and system condition assessment seem limited only by imagination and economics. We have to find economical ways to assess condition rapidly and reliably and to use information technology systems to manage the large quantities of information. Figure 5.2 shows the cumulative increase in available technologies.

USES OF CONDITION ASSESSMENT INFORMATION

Condition assessment information should be shared within and across organizations and used for multiple purposes:

- Operational evaluation of the need for preventive maintenance or repair
- Engineering evaluations of the need for renewal, including rehabilitation or replacement
- Budget information for the capital improvements program
- Accounting evaluation for GASB 34 purposes

These purposes are linked. Technical, planning, and accounting staff members use the information to perform remedial work, to make plans for capital improvements, or to maintain the utility's financial accounts. The planner/budgeter might ask the engineer, "What is the condition of the system?" The engineer might reply, "This section needs repair or replacement badly and ought to be included in the capital improvements program." The planner/budgeter will likely not ask "How do you know?" but might ask, "What is the priority of that project and how much will it cost?"

FRAMEWORK FOR CONDITION ASSESSMENT

Having a framework to explain condition assessment provides a common footing for discussion. Condition assessment is a familiar term, so people may take it for granted. However, even though leading utilities use condition assessment, others do not even perform it and still others do not understand it well.

Figure 5.3 illustrates the proposed framework for condition assessment. Following the work of Ellison (2001), the figure separates the condition of systems and the condition of pipelines. With the condition of components at the basic level, the conditions of components are then aggregated to arrive at the condition of systems. The other feature of the framework is the linking of condition assessment activities with the uses of the information. This framework uses the work of Deb, Hasit, and Grablutz (1995) to include performance information.

The framework has two types of indicators for pipe components, one related to the capital improvements program and one to operations. The capital indicator is infrastructure readiness or simply, "Will it last?" This indicator has two sources of information: (1) how secure the infrastructure is from damage from natural or human-caused threats; and (2) how robust it is against degradation, such as corrosion.

The operational indicator is operations readiness or "Will it provide capacity?" This has two basic dimensions: (1) the component delivers the *quantity* of water needed, and (2) it supplies the *quality* of water needed.

The supposition is that if the component is ready in terms of capital and operations, it has good condition or high integrity. Theoretically, if all components are ready, the system should be ready also.

In reality, it is often difficult to assess all components, so condition assessment methods must also focus on systems by aggregating information on components. Ellison's (2001) report explains how to evaluate system capacity and reliability as opposed to that of components. He points out that no single index can measure system readiness and suggests using break and leak frequency, unaccounted-for water (UAW), renewal rates, and renewal costs as measures of system renewal needs.

Condition assessment requires multiattribute measurement, and, ideally, a composite condition index could be compiled for each facility. However, there is no single best approach to compile such an index, and condition cannot normally be reduced to a single score, although

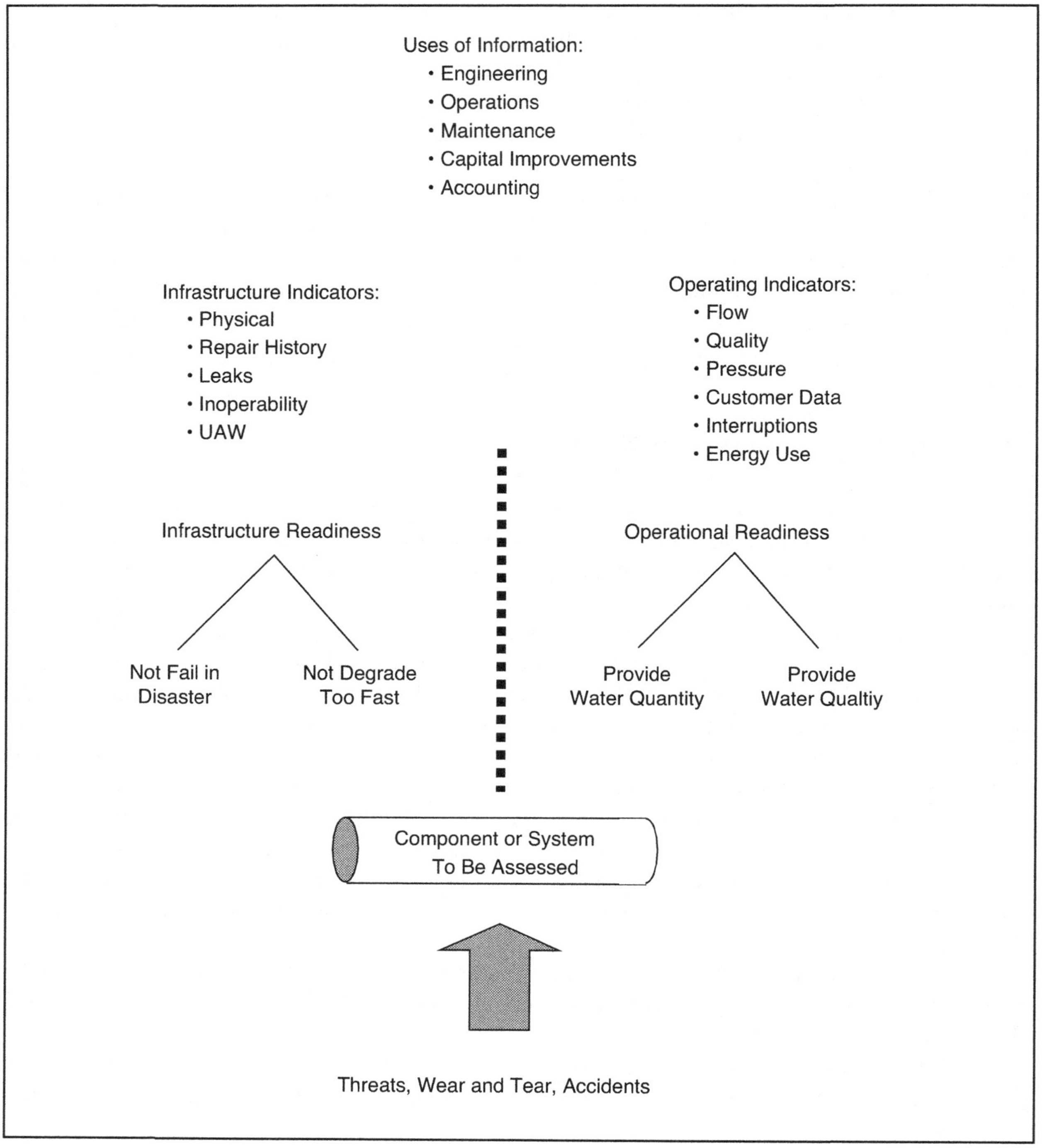

Figure 5.3 A framework for condition assessment of pipes and pipe systems

infrastructure "report cards" do that to emphasize the need for infrastructure investments. Hudson, Haas, and Uddin (1997) discuss methods of compiling composite condition indices. Ellison (2001) wrote in his AwwaRF report that single condition indices are not feasible for water distribution infrastructure.

Common indicators for condition assessment are wall thickness, cracks and fractures, broken wires, leaks and breaks, weak joints, tuberculation, and internal depositions.

Table 5.1
Variables that predict pipe failure

Materials and installation	Environmental conditions	Service conditions	Vulnerability
Poor original materials	Corrosive soils	Corrosive water	Digging
Defects	Ground movement	Water temperature	Trenchless intrusions
Age	Freezing	Traffic loads	Disasters
Type of material	Stray currents	System pressures	Cross-connections
Poor bedding		Water hammer	Nearby loadings
Weak joints		Service history	

AwwaRF research by Deb, Hasit, and Grablutz (1995) has offered insight into parameters to measure for condition and performance.

Indicators of the condition of components and systems should measure the variables that predict failure, as shown in Table 5.1.

FAILURE MECHANICS

Understanding the causes of failure in a system is important because systems are unique, and a solution technique might not transfer from one system to another. In assessing condition, we want to know the readiness of pipes to perform their functions; when systems or components cannot carry out these functions, a failure has occurred. Obviously, failure modes will depend on pipe type, installation conditions, pipe history, and other factors (O'Day et al. 1986). We identified a number of papers and reports on pipe failure investigations and have summarized a few here that are not mentioned elsewhere.

It is clear from the literature review that pipe failures happen through different modes. They may occur all at once or in multiple stages. They can be linked to a variety of circumstances, including human error or activity such as construction practices, routine service conditions, installation conditions, manufacturing defects, or accidents.

Hydraulic failure is the point at which a system becomes unable to provide required flow at a given pressure. Failure also may occur when physical events lead to unsatisfactory water quality. Additionally, pipes of all materials are susceptible to degradation or corrosion, and condition can be worsened by hydraulic conditions such as pressure surges and water hammer. The environment surrounding the pipe can also influence pipe condition. For example, the soil type can affect corrosivity, and soil displacement can literally pull apart pipe segments. Soil corrosivity and corrosion should be studied in conjunction with failure mechanisms. Corrosion advances steadily, but breaks resulting from corrosion will occur rapidly, once the first breaks occur. Temperature extremes can also be a means of failure. For example, many early CIPs were sealed with a leadite seal, and as a result of the different coefficient of thermal expansion between the pipe and seal, cracks may develop at extreme temperatures when the pipe expands more than the seal will allow.

Early on, most pipes were cast iron and they remain in place, so many papers have reported on factors causing them to fail. Morris's (1967) paper in *Journal AWWA* is cited often. It is one of the earliest comprehensive treatments of CIP failure causes. Other early papers were written by Arnold et al. (1960); Baracos, Hurst, and Legget (1955); Dolson (1955); and Wolfe (1946).

In his survey of utilities, Mays (1989) found that, generally speaking, cities did not maintain databases with information on networks, maintenance, failure, or repair. Nor did they have methods in place for optimal maintenance, rehabilitation, and replacement, according to Mays' survey. After reviewing other studies of the time, this picture seems a little pessimistic.

Allen (1985) discusses factors that contribute to failures and predictive models. Moruzzi (1986) discusses causes of main breaks and leakage and examines pipeline materials and resistance to failure. Male and Walski (1990), among others, discuss the various failure modes. Price (1990) examines the evaluation of prestressed concrete pipelines to locate damaged sections. Causes of pipeline deterioration and a probability model are discussed in a paper by Herz (1996). Six utilities representing diverse geographic, demographic, and geologic conditions were investigated in O'Day's study (O'Day et al. 1986), which is described elsewhere.

Kirmeyer, Richards, and Smith (1994) surveyed utilities to determine modes of pipe failure. Each report or survey obtains slightly different results, depending on the utilities that participate. This was true for the focus groups we convened in this project as well. They reported installation conditions, loads, service conditions, accidents, soil displacements, temperature extremes, and degradation of metal, concrete, or plastic pipe. Refer to Table 2.4 for a summary of data from the Kirmeyer, Richards, and Smith (1994) report, which provides typical causes of main breaks and can be used in failure diagnosis. See also the text by Smith et al. (2000).

Ferry (2002) wrote an article in *Plumbing Engineer* titled "Discovering Why Pipes Fail." Dingus, Haven, and Austin (2002) reported failure mechanisms of pipes 16 in. and smaller. They found, in order, corrosion of CIP, frost heave/ground motion, lead joint leaks, installation, A–C problems, and PVC pipe splits. Figure 5.4 shows a pipe deteriorated by tuberculation. For larger pipes 24 in. and greater, failures were caused by joint leaks/failures, damage by others, corrosion, fracture, and frost heave/ground motion. These vary by pipe type.

In Cleveland, beginning around 2000, a mayoral task force used Makar's failure classification system to investigate main breaks. Makar's system has three classes:

1. Class I (routine, day-to-day breaks, to be repaired as they occur; no major damage)
2. Class II (pipes in a region with high break rate)
3. Class III (serious consequences; many customers losing service; damage to surrounding environment; other direct, indirect, or social costs; expectation of review by managers, politicians, litigants, and media).

Cleveland's work provided excellent insight into ways and results of investigating failures of large mains.

Investigators outside the United States have also examined failure modes. A case study in northern Bohemia showed that failure rates and leakage increase in older mains despite maintenance efforts (Herz 1998). Results from a study of water pipe breaks in Malmo, Sweden, analyzed 734 breaks from 1985 to 1994 (Sundahl 1997). A report of a major pipeline failure in 1993 in Newport, Rhode Island, included problem identification (Anderson 1995). Matsushita's (2003) presentation described effects of the 1995 Kobe earthquake on the distribution system. Failures included two raw water conduits; six transmission mains; 1,757 distribution lines; and 89,584

Courtesy of City of Sioux Falls, S.D., Public Works Department

Figure 5.4 Deteriorated pipe caused by tuberculation

service connections. Cost of repairs for these items, including indirect costs, were about 25 billion yen ($200 million). Pipe materials in Kobe include cast iron, ductile iron, A–C (which has been replaced), and small quantities of steel and PVC. CIP failed both in pipe body and joint, while ductile-iron failures were mostly in joints. Seismic joints performed well.

Lessons learned from Kobe were that they require a redundant transmission system, CIP should be replaced with ductile and seismic joints, and that the distribution system is the key to quick recovery of water service after an earthquake. They planned a special seismic system using loops to recover water service quickly. For research needs, they see a need to identify vulnerable pipes and failure possibilities, learn how to prioritize replacements, and find out the ideal distribution system. Their planning method involves a database, which is updated with renewal information, pipe network analysis, engineering information, and management decision making about investments. They use a technique called a "raider chart" to assess blocks for hydraulics, water quality, aging, and seismic risk. They also factor in the cost of breaks to perform cost–benefit studies. They believe that more research is required for evaluation methods and cost–benefit analysis.

Obviously, the sheer number of different failure modes makes prediction difficult. However, it is important to record the causes of failure to forecast maintenance or replacement needs. For example, a break resulting from a construction accident would not help to predict age and corrosion failures. Therefore, data on the cause of leaks and breaks should be compiled and organized in a way that allows for accurate failure prediction.

This area of study can be called "failure mechanics" and "forensic analysis," among other terms, and it is closely related to the field of reliability engineering. Because reliability analysis requires failure data, its validity is usually limited by available data. Studies of distribution system reliability present analytical techniques, but the underlying data are usually not adequate to give confidence in the models. The availability of data and how they are used continue to be important research issues.

Reliability analysis is based on risk assessment. Pipes serving large areas are obviously of more concern than smaller local pipes. Ellison (2001) writes that the most common technique to identify critical pipes is a weighted-score approach, which produces a scorecard using such data as leak history, pipe size, pressure, age, damage potential, soil corrosivity, material type, and importance to the system. Sophisticated versions of databases for these data are available. Although the downside to this approach is the subjective nature of the scores, at least this technique provides a way to sort pipes into critical and noncritical groups.

In addition, quantifying the reliability of a water supply system requires "coping with aging and deteriorating infrastructures and prioritizing infrastructure improvements" (Harberg 1997). Harberg suggests identifying critical elements, usually in the transmission system (see also Goulter et al. 2000). See Cesario's text for further information on design criteria and reliability, and Mays' text for a discussion of reliability analysis (Cesario 1995, Mays 1989).

Many reliability concepts can be gleaned from other fields. Two useful concepts, with terms that are practically self-explanatory, are "fault tree analysis" and "failure modes and effects analysis." Bell Labs developed fault tree analysis, a primary method for performing reliability and safety analysis, in the early 1960s for defense work. Fault trees represent the interaction of failures and other events within a system; they can analyze large and complex systems and are particularly good for analyzing redundancy arrangements. In addition, common cause events are easily handled. In one paper we reviewed, fault tree analysis was used to estimate overall network reliability (Jacobs and Goulter 1991).

METHODS FOR PIPE CONDITION ASSESSMENT (COMPONENTS)

Assessing condition of a component means looking at an individual segment of pipe or part of a distribution system. Following Ellison's (2001) report, we discuss condition assessment for components and systems as separate topics. We recognize, of course, that system condition should be the aggregation of component condition. However, some methods, such as hydraulic testing, inherently assess a number of components.

Many of the evolving methods for component assessment rely on NDE or nondestructive testing (NDT). These terms describe methods of inspecting or testing an object without altering its characteristics. Today, NDE is widely used for industrial applications, but it is just beginning to be used in evaluating water pipelines.

Much of the progress in NDE came after catastrophic failures. Today's NDE began in the 1920s and advanced during and after World War II. In 1927–1928, a train wreck prompted the development of a technique using electric current induction/magnetic field detection. In 1929, methods using electromagnetic conduction and induction came into play. In the 1930s, gamma radiography was used, and later, advancements were made in penetrant testing, eddy current testing, and ultrasonic methods. In 1941, the American Industrial Radium and X-Ray Society was formed; this later became the American Society for Nondestructive Testing. In the 1950s and 1960s, acoustic emission testing and thermal imagery were introduced (Hellier 2001).

Reports of NDE techniques contain descriptions of their application to water mains. For example, Brown (1998) includes articles on electromagnetic acoustic transducers, digital radiography of process piping, ultrasonic detection of pipeline cracks, magnetostrictive sensor technology for defect characterization, interferometric synthetic aperture radar technology to detect ground movement, eddy current testing, thermographic leak detection, in-service pressure vessel inspections, flexible mobile robots for smart optical pipe inspection, profile measurement for wall thickness, electrochemical corrosion sensors, fiber-optic chemical sensors for corrosion, impulse-response systems, GPR, and others.

Locating Pipes

Locating and gaining access to pipe is the first part of condition assessment. Deb et al. (2001) summarized the state of the art in locating buried infrastructure. AwwaRF and U.K. Water Research Ltd. (2002) held a workshop on "Multi-Utility Buried Pipes and Appurtenances Location." Another AwwaRF report that includes information on pipe location is by Ellison (2001), who gives information that is mostly from Von Huben (1999b). Individual cases, such as DeKalb County, Ga. (Buie 2000), are also reported to explain specific experiences. DeKalb County's global-positioning system (GPS)/GIS/Data Management Department uses pipe locators, dowsing rods, metal detectors, potholing, and probing, and its GPR helps in locating nonferrous pipes.

Deb et al. (2001) surveyed several water and gas utilities and learned that most used magnetic and electromagnetic equipment. A few used S&A equipment. For research needs, the utilities reported that they need equipment to effectively locate and identify nonmetallic assets, equipment that will not be affected by other utilities, equipment to distinguish gas pipes from water pipes, and equipment to accurately determine depth of assets. The team also further evaluated five technologies, including magnetic, EM, GPR, S&A, and infrared thermography. These investigators field-tested GPR, EM, and S&A for vertical accuracy, horizontal accuracy, and limitations of each technology. They describe different models within each group and list the pros and cons. This report, then, contains useful information on how well these technologies actually work.

The workshop sponsored by AwwaRF and U.K. Water Research Ltd. (2002) focused on location techniques, but it also included discussions of asset recording and condition monitoring. It discussed not only water, but also gas, fiber optics, electricity, oil, chemicals, cable television, telephone, sewers, and other underground structures. Project workshop participants began by acknowledging that better data processing equipment was needed and that records get lost. Attendees considered GPR, electromagnetic line locators, magnetometers, acoustic locators, infrared thermography, and underground mapping, and compiled research needs in 12 areas divided into "drivers" and "blockers": philosophy, regulatory needs, strategic thinking, economics and social need, standards and specifications, capturing and recording data, proactive measures, finding small things, equipment performance, data sharing, technological shortcomings, and training.

At the conclusion of the project workshop, the participants developed 16 proposals for research:

1. Methodology and standards for utility asset location data collection and exchange
2. Smart technology
3. Methodology for determining total costs for construction and O&M of buried infrastructure
4. Asset tagging
5. Development of a multisystem location system
6. Improved GPR performance
7. Smart pigs for location of adjacent buried infrastructure
8. Feasibility of using horizontal sensing to determine depth of buried infrastructure

9. Development of "see ahead" technology
10. Novel approaches to traditional underground infrastructure
11. New technologies for underground asset location—capability review
12. Quality training
13. Requirements analysis of user needs
14. Movement toward a better regulatory framework for buried infrastructure management
15. Techniques to detect existing nonferrous buried assets
16. Utility environment categorization

NDE Techniques

AwwaRF was studying NDE by the time its study, which reviewed the field, was published in 1992 (Jackson, Pitt, and Skabo 1992). These investigators studied ultrasonic testing, acoustic emission, MFL, and RFEC, with a focus on CIP, DIP, and RFEC. They did not study CPP but suggested that low-frequency ultrasonic and acoustic emission methods might be useful in situ. They also did not study steel because technology is available from the oil and gas industry, and steel is a low percentage of water piping. Finally, they did not examine plastic and A–C pipe.

Another report by the Construction Industry Research & Information Association (CIRIA 1996) deals with environmental and service conditions or inspecting pipe after excavation. In situ methods analyzed were closed-circuit television, fiberscopes, and hammer testing. The report concluded that only visual methods could be used in situ but that RFEC, MFL, and ultrasonic testing offered promise for metal pipes. It also described the WSSC method for acoustic studies of CPP.

Dingus, Haven, and Austin (2002) produced a more recent comprehensive study for AwwaRF. The goal of the project was to "evaluate the feasibility of developing technologies that can be used by water utilities to assess the condition of buried pipelines without excavation and coupon sampling." The project was to "select and field screen potential technologies for pipeline condition assessment." The team carried out literature searches, a survey of utilities, and some testing at utilities. The Nondestructive Testing Information Analysis Center supplied literature searches and vendor data. From the literature review, these authors reported that NDE methods are used in other fields but have not been transferred to water very intensively. The literature review showed that no purely aboveground methods are available and that inspection methods rely on a pig to gain access to the pipe.

Dingus, Haven, and Austin (2002) surveyed 100 large water utilities and received 46 responses. Questions were on failures, inspection practices, and replacement guidelines for NDE. Forty-six percent of the reporting utilities had an inspection program, and all said that they use visual inspection techniques. Eighty-one percent have a replacement program, and most use failure rate as the basis for replacement, but 27% use additional data, such as statistical prediction, soil analysis, and repair costs. Utilities replacing a significant amount of pipe were replacing CIP with another type. Of the responding utilities, 61% had not considered using NDE. Other points working against NDE were low experience level, few NDE vendors, and big improvements needed in concrete.

The report describes testing technologies and methodologies, along with field tests and results. In their report, these investigators point out that visual inspection and hammer testing are currently the most common methods. Pigs, data logging, and NDE sensors need to be developed and optimized for water utility applications. The report presents a table listing inspection methods, pipe material, defect types, and notes, and the authors describe the difficulty in using pigs from the oil and gas field (because valves, turns, and entrance and exit points present problems). Specifically, they found that no method exists for polymeric pipe, RFEC is the most accurate and advanced method for metallic pipe, and NDE for concrete is steadily advancing. Results of field tests could not be totally verified because it was too expensive to destroy pipes to check results. The Dingus, Haven, and Austin (2002) report also includes approximate costs for NDE pipeline inspection. Barriers to NDE were access, pipe disturbance during inspection, service interruption, water quality issues, cost, effectiveness, and accuracy.

Pigs are usually used in long runs of pipes without tees, elbows, joints and unions, reductions, and valves. Articulation technology from boiler tubes that can traverse through 180-degree bends might be used in water applications and could be studied by research. Also, it is hard to get NDE equipment into water pipes because launch and catcher systems are lacking.

Research needs are to find ways to inspect pipe of the future, especially polymer-type pipes. Also, insertion methods and the ability to pass bends are needed. Smaller hydrophones for acoustic emission are needed. Other needs are for piggable (fully opening) valves and RFEC for concrete-reinforced pipes.

Visual Methods Using Video

Closed-circuit television surveys are, generally speaking, not as useful in water pipes as in sewers. Exceptions might be large pipes and situations where deteriorated cement lining can be found (Ellison 2001).

Electronic—RFEC

RFEC is a useful technique for metallic pipes (Jackson, Pitt, and Skabo 1992; Dingus, Haven, and Austin 2002). The technology goes back to the discovery of electromagnetism, and it has become an important method for testing materials for discontinuities. Although it is a valuable NDE method, the test material must conduct electricity. Also, eddy current testing is mainly effective for discontinuities at or near the material surface.

The RFEC technique had been used to assess defects in well casings and in boiler tubes. In pipes, it uses a "smart" pig that impresses an electromagnetic field on the pipe wall. A remote detection coil senses the signal and relays it to a computer for analysis. Signal analysis yields data on corrosion, pitting, defects, graphitization, and thinning.

Steel pipe is more difficult because of thinner walls and irregularities such as rivets. It is a candidate for controlled destructive evaluation (CDE), which the LADWP has adapted for a large-scale program (Ellison 2001).

RFEC will not work on nonmetallic pipes such as PVC, high-density polyethylene (HDPE), and A–C, but it is being adapted for prestressed concrete. Mergelas and Kong (2001) describe how RFEC/TC (Remote Field Eddy Current Transformer Coupling) detects broken wires in a PCCP. The technique can identify and rank wires by severity of wire breakage.

The Mergelas and Kong report (2001) presents the pros and cons of current inspection methods and field tests on RFEC/TC on large-diameter pipes (greater than 48 in.) Limitations, including that the restriction of the method by calibration, are pointed out. An increase in accuracy is seen with an increase in the number of breaks, and there is a high confidence in the method that ranks the damaged pipes. The pipes need to be free of water, and measurements are taken at a walking pace. The pipes are then ranked by order of condition. This method gives a distribution of distress, classifying by the degree of distress, and can predict the number of broken wires when calibrated (gives a rank when not calibrated). It gives no information on the circumferential position of the broken wires, and breaking at the end is harder to detect. Overall, the method allows the manager to make informed decisions and is thought to be accurate.

Acoustics and Acoustic Emissions

When a stress is placed on an object, discontinuities within the object release energy in the form of high-frequency stress waves, which can be interpreted according to their frequency and intensity. A transducer can convert these waves into a voltage. We can hear the sound associated with plastic deformation at the tip of a crack before propagation (Physical Acoustics Corporation 2003).

Acoustic emission testing can be performed while a component is in service, yielding real-time data. The test detects movement instead of discontinuities. Considering its performance, this technique is also relatively inexpensive. Nevertheless, acoustic emission is limited in its repeatability because results are subject to a particular loading scenario. Also, the test is subject to interference from other noise sources. Acoustic testing has been used in the pipeline industry (e.g., in PCCP testing; Holley 1999). Acoustics are also used in leak detection.

Ultrasonic Testing

Ultrasonic testing can be used to assess pipelines after installation. In it, high-frequency sound waves are used to detect discontinuities in materials. The waves are projected through a material but are reflected at interfaces such as the back surface or internal defects. The reflected waves are read by a transducer, which converts sound waves to electrical energy. Ultrasonic testing detects small, subsurface discontinuities and can determine the depth of the discontinuity. Furthermore, the equipment is relatively light and portable, and no special user licenses are required. Limitations occur because defects parallel to the direction of the waves, or similar in structure to the material, may not be detected. Results also may be difficult to interpret, requiring considerable training before they can be read accurately. Ultrasonic testing has been used for a variety of applications, especially in metal structures (Mech-Chem Associates 2002).

Infrared and Thermography

Thermography is relatively new compared to other NDE technologies. In thermal infrared testing, the surface of a test material is heated and then monitored with an infrared camera. Discontinuities in the object will interfere with the normal flow of heat, appearing hotter than the rest of the material. A limitation is that only the surface of the material can be viewed. It holds promise for the future, but accuracy depends on skill and experience. In addition, many factors affect the thermograph, such as emissivity of the material, ambient air temperature, and heat-transfer characteristics.

Infrared scans of electrical equipment and water utility facilities can pinpoint such problems as electrical shorts, heat escape, and unwanted air filtration. Thermography is being utilized at water treatment facilities in Mount Clemens, Grosse Pointe Farms, and Ann Arbor, Mich.

Other NDE Methods

Liquid penetrant testing can indicate discontinuities in a material by applying a liquid penetrant to a nonporous material. It works by capillary action where surface tension draws liquid into small openings and has the capability to detect manufacturing flaws.

Magnetic particle testing is relatively simple to use and interpret. The particle indicators represent the size and shape of defects, and cracks can be detected even if filled with a foreign material. However, the technique has drawbacks. Because it can be used only for ferromagnetic materials, it would not be useful for plastic pipes. Also, it is only effective in detecting defects at or near the surface. Performing this procedure inside a pipe may not detect discontinuities on the outside pipe surface. As a result, this technique may be useful in pipeline evaluation only before the pipe is installed (Hellier 2001).

Radiographic methods are sensitive and accurate and have been applied to the evaluation of welds and casts. They would be most practical to use to test pipe for manufacturing defects but are difficult to interpret.

Magnetic resonance imaging (MRI) has been used primarily in the medical field and is a form of NDE for living organisms. An MRI scanner has a large semiconducting magnet that produces a magnetic field with an intensity on the order of 20,000 times the intensity of the earth's magnetic field. In addition, the MRI scanner emits a radio frequency pulse that will cause hydrogen atoms to absorb energy and spin at a particular frequency in a particular direction. At the same time, three gradient magnets are turned on and off rapidly. As a result, the image will be localized. MRI scanners are currently very large and expensive, and the subject is required to be very still. Perhaps similar technology can someday be used for assessing pipeline condition (Gould 2003).

Physical Sampling by Coupons

Spot condition assessment and coupon testing are not normally advisable, according to Ellison (2001), except in special cases that might include A–C pipe softening (which is usually uniform), or to calibrate an RFEC survey. Although in situ assessment is changing rapidly, it is still expensive and difficult, and disinfection is an issue (Ellison 2001).

Controlled Destructive Evaluation

CDE offers a direct way to test pipe integrity. The LADWP has adopted CDE for a large-scale program.

METHODS FOR DISTRIBUTION SYSTEMS ASSESSMENT (SYSTEMS)

Deb, Hasit, and Grablutz's (1995) report on performance evaluation is a good overall discussion of condition and performance evaluation. Its goals included guidelines to evaluate the condition of distribution systems and establish target levels of performance.

Figure 5.5 Leak detection

Water Loss Accounting

Work on water loss accounting and control has proceeded (see Figure 5.5), with much of the activity taking place through AWWA's Water Loss Control Committee (formerly the Leak Detection and Water Accountability Committee). According to its Web page (AWWA 2004), the committee's purpose is

> To increase water utility awareness of the nature and extent of water loss in the industry and improve the level of water accountability employed by water suppliers; by furthering the science and application of water accounting, leakage and pressure management and universal metering systems.

The committee issued a report in 2002, which was published in *Journal AWWA* in August 2003 (AWWA Water Loss Control Committee 2003). The report was issued while the project was under way but could not be cited until it was published. The report offers new thinking about resolving old issues, such as the lack of standards and measurement techniques for concepts such as UAW. In particular, the report summarizes the International Water Association (IWA) task force's report on a water audit methodology. It clarifies terminology, advocates standardized approaches to water accounting and loss control, and describes an international approach for water auditing and performance measurement.

A utility's leak detection program offers an opportunity to improve the database while solving leak and breakage problems and to organize data for main replacement decisions. Active work on leak detection and water loss control goes back more than 20 years. A number of papers have been published in *Journal AWWA*, including Siedler's (1982) early paper "Obtaining an Analytical Grasp of a Water Distribution System." AwwaRF published a report on the topic in 1987 (Wallace 1987), and later papers include one by Buie (2000) and numerous AWWA committee publications.

AWWA's Leak Detection and Water Accountability Committee reported in 1996 that the target for UAW should be 10% or less. The committee compiled a report in 2002 to advocate new international methods for water auditing and loss control (AWWA Water Loss Control Committee 2003).

Meanwhile, other authors presented case studies and other data (Nikonow 2001). In this reference, the author presents a method used to quantify leakage in the system of the Greater Johnstown Water Authority in Johnstown, Pa.

Thornton (2002), who was active on AWWA committees, compiled an extensive manual for water loss control. It covers the main technical subjects, such as water auditing, and includes a review of articles and case studies from North America and Brazil. The book also covers the United Kingdom experience.

In the late 1990s, AwwaRF published a report on leak detection in plastic pipes (Hunaidi et al. 1999). The goal of these authors was to investigate commonly used acoustic leak detection equipment—in particular, leak noise correlators—for locating leaks in plastic pipes. Field tests were performed and teams came in to blindly locate leaks in a test pipe. Findings were that operators were not inclined to measure the actual sound propagation velocity and relied instead on values already programmed in the correlators for different pipe types and sizes. When the measured propagation velocity was used, leaks were pinpointed accurately. Tests using a laboratory-grade state-of-the-art vibration measurement and analysis system were carried out. Three technologies were investigated from other industries: GPR, thermography, and tracer gas. In short, commercial modern leak noise correlators were generally found to be capable of locating leaks in the test PVC pipe; however, several improvements could be incorporated into existing equipment and field procedures to increase their effectiveness.

Farley and Trow (2003) explain water loss control as it is practiced in the United Kingdom. Substantial progress has been made, as monitored by the U.K.'s Office of Water Services.

P. Fanner (2003) showed how leakage was reduced in England and Wales using a mixture of measures, including component analysis, district metering, repair and replacement programs, and pressure reduction. Component analysis is a set of concepts for estimating, managing, and regulating, and is based on U.K. research. Fanner cited work of an IWA task force that has become AWWA policy.

Pigs

Other types of smart pigs based on ultrasound and MFL have not worked well for water pipes (see Jackson, Pitt, and Skabo 1992; Ellison 2001).

Courtesy of City of Sioux Falls, S.D., Public Works Department

Figure 5.6 "C" factor flow test

Hydraulics

Hydraulic capacity is measured by a comprehensive capacity evaluation using system data and models (see Figure 5.6). Ellison cites Cesario's (1995) text as explaining this process well (Ellison 2001).

DATA SYSTEMS FOR REPORTING CONDITION OF COMPONENTS AND SYSTEMS

Papers on data systems show the need for work on using data to predict pipe failures. They also show the increase in knowledge about pipe failures and problems. If the data can be managed effectively, a great deal can be learned. Kleiner and Balvant's (1999) work discusses how data on pipe classes can be partitioned to predict failures. Their approach shows how far we have to go before data can actually be used to prioritize on a proactive basis. These authors point out that much more data is required before a single pipe analytical approach, such as that described by Andreou, Marks, and Clark (1987), can be used.

Taken together, utilities have substantial information on condition of their systems. For example, WSSC has compiled a great deal of information on the condition of its system. This information is included in internal reports such as *Water Main Condition Assessment, 1915–1998, Analysis and Recommendations;* and *Water Distribution System: A Collection of Statistics, Facts, and Comments* (WSSC 2002).

Although they are not fully documented in published papers, WSSC's practices illustrate a thorough approach to using system data to track the condition of water mains, compile the information into databases, and use it for planning of renewal projects. WSSC's databases actually include four computer-based applications: the maintenance management information system (MMIS), mapping, GIS, and CAD. The MMIS is linked to work programs and contains database information from work orders, contracts, and repairs. It was programmed in the 1980s and WSSC is planning a conversion to a new platform. The GIS allows manipulation of spatial data, including many of the features of Webmap. The CAD package is used to prepare design drawings and allows importation of information from the other packages (Woodcock 2003).

Deb et al. (2002a) also discuss main break databases. In their survey, 70% of North American and 85% of European utilities indicated that they had computerized records of main failures (however, this does not reveal how well the data are managed or maintained). These authors believe that useful applications of data can be made, even with limited data. They recommended a comprehensive program of main break data collection consisting of field, office, testing, and cost data. Recommendations were for all utilities to collect and maintain water main break data, and guidelines were presented that include 45 data items to be collected for a main break. This illustrates the complexity of the problem.

CORROSION SCIENCE AND CONTROL

Much of condition assessment deals with corrosion and corrosion control programs. This is discussed briefly here and in more detail in chapter 7, where cathodic protection is presented as a rehabilitation technique. Corrosion is a broad subject, and we provide only an overview and a brief discussion of research needs.

AwwaRF has published a number of reports about corrosion, which include those by Reed and Tucker (1988); AwwaRF and DVGW-Technologiezentrum Wasser (1996); Reiber et al. (1997); and Edwards, Reiber, and Schecher (1997). As a result of these and other studies, a great deal is known about internal corrosion, but it remains a very complex process that affects integrity of pipes. Perhaps less is known about external corrosion than about internal corrosion, and AwwaRF has just funded a new report to summarize the knowledge of the external phenomenon.

A recent AwwaRF study by Rajani et al. (2000) presented information on optimum time to replace gray cast-iron water mains after considering corrosion. Several technologies were investigated. The most promising were found to be ultrasound and RFEC. All field tests were done on 6- and 8-in. pipe. The results were that ultrasonic inspection is simpler but will not work where tuberculation is present. This means that RFEC will be a more flexible method to size defects in three dimensions.

This study examined external corrosion to estimate pit depth and soil properties characterized by soil pH, resistivity and soil aeration. The authors encourage utilities to develop databases and take samples during the repair and replacement of water mains. The report reviews mechanical properties of gray cast-iron mains, mechanical tests (with photos), metallurgical analysis, soil as related to corrosion, and NDT.

McNeill and Edwards (2001) reviewed almost 300 articles that embrace 100 years of water utility experience with corrosion. Factors they studied include water quality and composition, disinfectant residuals, pipe age, scale formation, temperature, flow conditions, biological activity, and corrosion inhibitors. The authors review how proposed regulations such as the LCR, the Disinfectants/Disinfection By-Products Rule, and others may affect iron pipe corrosion. To summarize their findings, the effects are apt to be mixed. The authors write "Iron pipe corrosion is extremely complicated and is affected by practically every physical, chemical, and biological parameter in water distribution systems." The work is concerned with interactions between the water and metal and does not address external corrosion; it could be retitled so as not to imply that it covers all aspects of corrosion.

Water supply industry representatives believe that the industry should modernize with regard to internal corrosion control. In a volume on internal corrosion, an international team of authors concluded that "...the water supply community has failed to keep pace with corrosion science and has only recently begun to import corrosion technologies developed in other industries." These authors thought that methodological and assessment protocols should be advanced in several areas (AwwaRF and DVGW-Technologiezentrum Wasser 1996).

In addition, these investigators recommended using standardized coupon exposure procedures that will yield information on corrosion rates, surface scales, and pitting propensity. They thought that a database on materials corrosion and water quality should be organized. Finally, they recommended standards for the electrochemical measures of the reactions that drive corrosion.

External corrosion can be affected by soil resistivity, pH, and other factors. The authors used a GIS to correlate soil data and external pipe corrosion data, showing that mapping is a useful tool in such corrosion studies (Doyle and Grabinsky 2003). Figure 5.7 shows repair of a pipe subject to corrosion failure.

Reiber et al. (1997) examine decision making at water utilities attempting to mitigate lead and copper corrosion by-product release. This work is also a technical manual for the CORRODE software.

Reed and Tucker (1988) discuss evaluation of a pilot versus field distribution corrosion monitoring system. A technical explanation of corrosion rates and corrosion-inhibiting compounds is given. Both laboratory and field tests were described and performed, and the results are presented.

An important issue that possibly requires more research is whether polywrapped ductile iron is as corrosion-resistant as pipe with bonded coatings (T. Johnson, U.S. Bureau of Reclamation, personal communication, Aug. 7, 2003).

WATER QUALITY AND CONTAMINANT MONITORING

We did not review distribution water quality monitoring in the project, but water quality is mentioned briefly in this chapter because it is, in some ways, an assessment method. Water quality monitoring can be done for regular operations or to detect intrusions.

At the project workshop, Troutt (2003) presented a paper on water quality monitoring as it is practiced by Tampa Bay Water (Florida), which is the largest wholesale water provider in the southeastern United States. Tampa Bay has a four-step plan to ensure infrastructure integrity: the water quality monitoring plan, continuous water quality monitoring at each treatment plant effluent point, continuous water quality monitoring at points of connection, and online (supervisory control and data acquisition) monitoring of operational parameters. The goal is to ensure that the regional water quality will not cause damage to infrastructure or customer infrastructure. Their monitoring program includes ongoing corrosion monitoring.

Pikus (2003) recently discussed approaches to monitor for contaminant intrusion and gave as its goal to detecting, identifying, and measuring malevolent contamination events that threaten significant public health risks in time to provide the support to initiate responses that will minimize or eliminate public health risks with an acceptable low rate of false alarms.

This author discussed classes of sensors that include sentinel species such as bacteria or algae, electrical and optical properties, and surface properties such as adhesion, biological assays, and chemical properties. Because so many key design issues must be addressed, Pikus concluded

Courtesy of Northern Colorado Water Conservancy District

Figure 5.7 PCCP pipe repair

that monitoring contamination intrusion is a complex problem that requires many trade-offs. Research needs include guidance in terms of contaminants of concern, type of sensors, siting parameters, data processing principles, and ways to control false positives and negatives. Research might identify a combination of condition and contaminant monitoring that utilities could implement.

Large required investments loom over utilities, but they are also concerned about water quality and regulatory compliance. This issue may perhaps be larger than infrastructure integrity as a future driver of needed investment. At the 2003 AWWA national conference, Singer (2003) identified this as a major future concern and reported that many unknown water quality reactions occur in distribution systems. Clark and Grayman (1998) summarized the issues in an AWWA text on water quality modeling in distribution systems and suggested that distribution failures can cause intestinal illness and deaths.

Ways that distribution systems influence water quality include

- Open reservoirs
- Enclosed, unchlorinated reservoirs
- New construction that disturbs the distribution system
- Main breaks
- Backpressure
- Dead ends in mains
- Living organisms that protect and release bacteria
- Sewage cross-connections
- Corrosion by-products
- Disinfection by-products (DBPs)

Research needs for water quality in distribution systems are summarized in chapter 9. They include (Clark and Grayman 1998)

- Model structure and technology, and guidance for model usage
- Influence of tanks on water quality and improved tank designs
- Legal and management implications
- System sampling and monitoring strategies
- Chemical, physical, and biological issues in distribution systems
- Disinfectant and corrosion by-products

PROJECT WORKSHOP EXAMPLES AND REPORT

Condition monitoring was discussed extensively at the project workshop, including specific cases such as Cleveland's (discussed under failure mechanics) and as a part of overall planning programs (discussed under planning and prioritization).

Woodcock (2003, 2002b) described WSSC's condition assessment program. This program takes multiple views of condition assessment that consider different materials, design components, and NDE tools to formulate an assessment process. This process comprises design, right of way risk assessment, inspection, and maintenance, along with operations history, materials, failure modes, and pipe specifications. Woodcock presented insights into specific failure modes for different types of pipe, including CIP, DIP, and PCCP. He concluded that technology to improve condition assessment does exist, but it is not being developed to any extent for water and wastewater. The industry needs to do a better job of stressing its needs to create incentives for developers.

Several participants focused on condition of PCCP. Galleher (2003) explained the San Diego County Water Authority's aqueduct protection program. The program uses internal inspections and RFEC, and it is designed to avoid catastrophic failure; protect structural integrity of existing pipelines; determine remaining service life and condition; extend pipeline service life; and develop a maintenance, rehabilitation, and replacement strategy. Procedures performed with empty pipe include sounding, RFEC, and impact echo. With the pipe in service they perform acoustical monitoring.

Zimbelman (2003) explained a similar program being implemented by the Northern Colorado Water Conservancy District. There, the Colorado–Big Thompson and Windy Gap Projects include a Western Slope collection system, a tunnel under the Continental Divide, an Eastern Slope distribution system, and a 6-mi long, prestressed concrete pipeline, 9 ft in diameter. The pipeline is operational during the spring and summer runoff season (i.e., May, June, July, and rarely, August). The district also used electromagnetic (eddy current) testing and identified several sections with potential broken wires and has completed a remediation program that included a new cathodic protection system.

At the project workshop, the working group on condition assessment and failure mechanics noted the value of condition assessment. Much progress has been made in U.S. systems in the last decade, with major advances that include the ability to analyze pipes without digging them up, pigging, depth and location finding, pressure testing, and CDE. The group thought that Europeans do a good job and that we should research differences between management systems.

Needs identified during the project workshop are listed here:

- Tools (x-ray methods; tools for all ranges of sizes, especially tools to see through concrete pipe and detect cracks; tools for tight bends, valves, and reducers; a smart tool to be sent down the pipe and retrieved with data that can be downloaded; small chip sets in pipes to give type and date of installation, for example, by inserting a "ball" into the pipe; smart pigs; encrypted identification; sensors to record sounds of breaks; fiber optics to record breaks in light; improved metering capabilities to narrow sources of leaks; better acoustic monitoring; RFECs patterned on those used in the oil and gas industry; MFL tools; real-time assessment methods; and better GPS)
- Good drawings and information (manufacturers, dates, software to organize the data, a national database for condition assessment information)
- Workforce needs (less labor-intensive methods of locating water main breaks, documentation of institutional memory of employees, training and workforce development, training with soil tests, training for software such as KANEW, and incentives for the private sector)
- Organization of a research foundation similar to that serving the oil and gas industry

Although a number of technologies are available, effective condition assessment tools for the various pipe types are lacking. Some of the obstacles to effective condition assessment are lack of records, inability to inspect pipe while in service, and lack of funding support by local boards and councils.

Options for overcoming the obstacles are limited, but generally, they are

- Developing better condition assessment tools, especially for specific applications
- Providing support and education for the programs
- Furnishing adequate resources and staff
- Educating management about condition assessment

Large utilities must be willing to take the risks to implement condition assessment programs so that small utilities will follow. All utilities must be willing to invest in research programs to learn the status of their underground assets. This will lead to awareness and the ability to inform other large utilities of the advantages of research. Utilities should be willing to share their wealth of information. Each utility will need a pipe failure database that includes specific location and other inventory data, such as size, depth, corrosion indication, type of break, repair methods and histories, and pictures.

Undertaking a utility research and action program will require

- A template to follow
- Incentives for gathering, collecting, and importing data
- Accurate and improved tools and materials

CHAPTER 6
PLANNING AND PRIORITIZATION

CHAPTER OVERVIEW AND PURPOSE

The purpose of this chapter is to review knowledge about prioritization of pipe renewal projects, as it fits into the framework of planning and management methods. In a nutshell, sophisticated methods for capital improvement planning and budgeting and powerful computer-based techniques are available to utilities, but in the end, if their decision making is to be effective, they need to know more about the actual condition of buried individual pipes. In addition, direct economic analysis will not be enough to assess the costs and benefits of water main projects; other social and management factors must also be considered (AWWSC 2002).

EVOLUTION OF PLANNING AND MANAGEMENT METHODS

Planning and prioritization belong to the field of management and its branches. Management tools from business and public administration can be used in any field where capital investments are required. We review the field briefly with Figure 6.1, which shows the growth of management subfields. Modern human resources refers to developing the basic competence of all workers engaged in the tasks. Law relates to all issues of contracting, liability and so forth. Public involvement shows how to consider criteria such as customer satisfaction. Economics and management science merge into the tasks related to decision analysis, which is used in prioritization. Finance is the home field for the budget process, which adopts the priorities set for the capital budget. Organizational development provides a framework for how sections of the utility work together to prepare plans and priorities. Computer hardware refers to the many mechanical innovations available, and software refers to how methods and processes are carried out by computation and communication.

DECISION TREE FOR PLANNING AND PRIORITIZATION

Figure 6.2 shows the decision tree for planning and prioritization. The tree has two basic branches: the management of the information needed and the process of planning and prioritization.

PLANNING AND PRIORITIZATION PROCESS

Each utility has a unique mixture of pipe materials and characteristics based on its development, construction practices, maintenance history, and soil conditions. However, a basic strategy of planning can be outlined, including priority setting. The purpose of planning and prioritization is to select projects for capital improvement, and funds must be allocated in competition with other utility needs.

The discussion begins with a review of the capital planning and budgeting process. Next, we turn to the elements of the process, including prioritization. The elements involved are how to analyze the pipe needs, which pipes to renew, how much to spend, priorities for renewal, methods for renewal, schedule for renewal, funding sources, and how to present the information to decision makers.

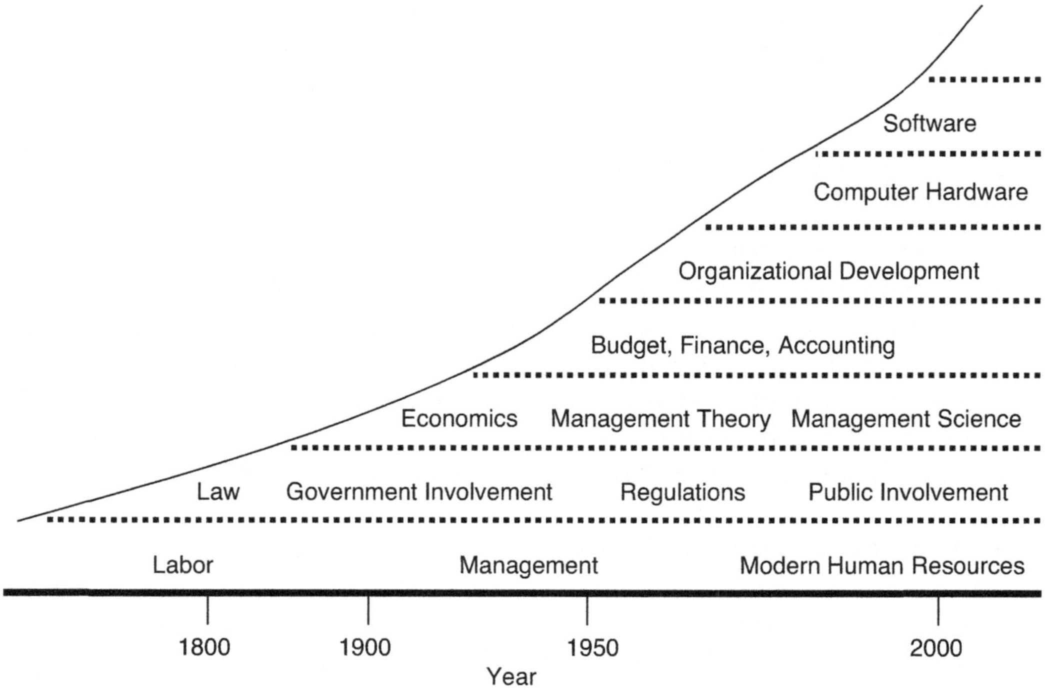

Note: The diagram is conceptual. Not all fields or years are shown.

Figure 6.1 Evolution of management practices

In the private sector, capital allocation models focus on return on investment and on risk allocation. For example, a capital asset pricing model has been used in capital budgeting to estimate an investor's required rate of return on equity (Ross, Westerfield, and Jordan 2000). However, most discussions of prioritization relate to public sector issues and include social as well as infrastructure costs.

Public Sector Capital Planning and Budgeting Process

Capital planning and budgeting have been evolving since municipal budgeting began early in the 20th century (Moak and Hillhouse 1975). The goal, then and now, is to determine how a potential investment rates in comparison with other choices.

The prioritization process for pipes involves comparing renewal projects with other types of projects (apples with oranges), and comparing competing renewal projects (apples with apples).

Therefore, synthesizing knowledge about prioritization includes planning, budgeting, performance assessment, optimization, and public involvement. These topics, which are all interdisciplinary and involve several fields of knowledge, fit into the larger field of management science, where prioritization belongs. Examples include business administration (capital budgeting), public administration (planning and the budget process), economics (performance evaluation), industrial management (optimization), political science (politics of budgeting), engineering (systems analysis and design), and utility management (capital budgeting for utilities).

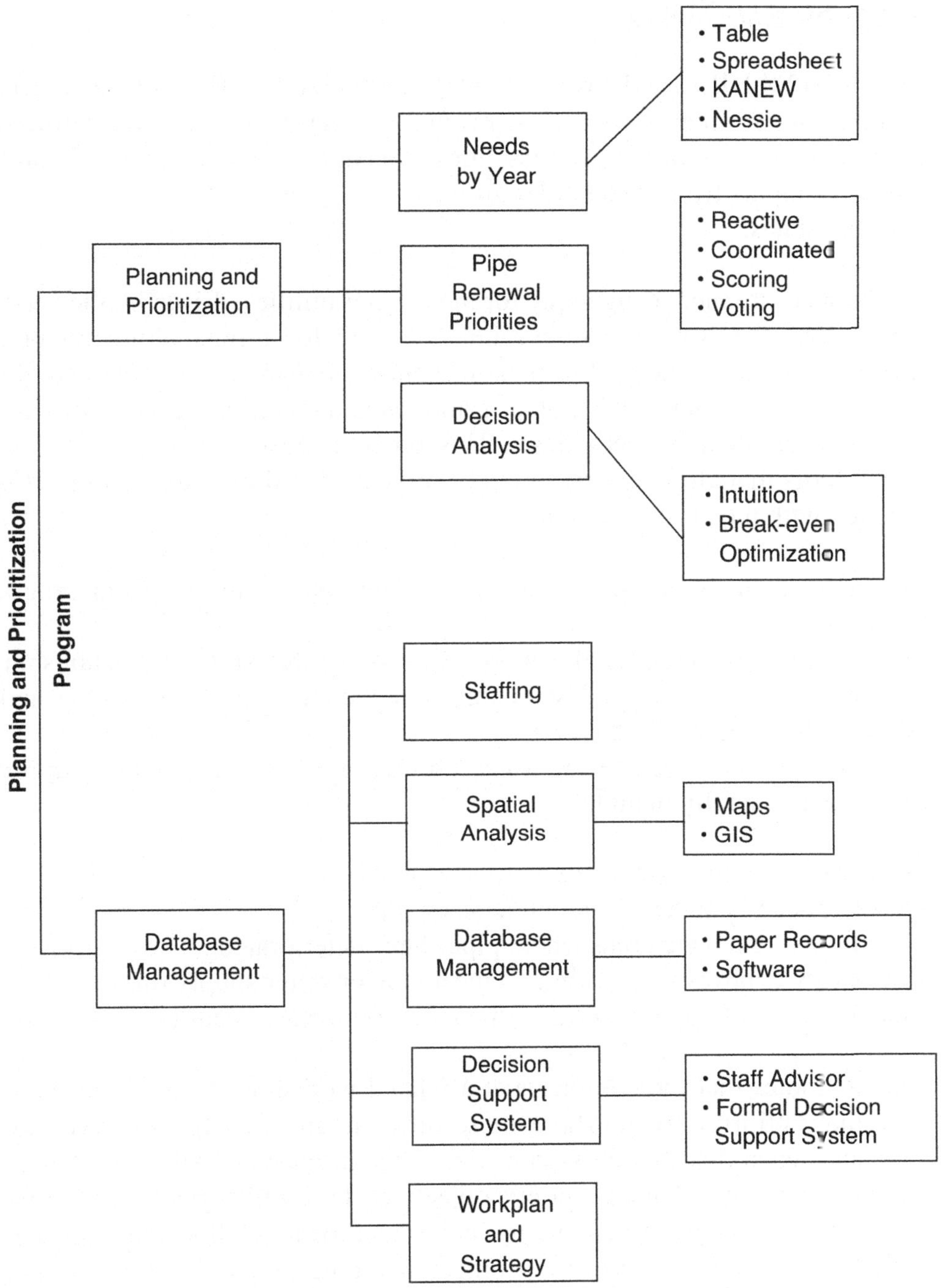

Figure 6.2 Decision tree for planning and prioritization

Capital Planning Strategy Manual

AwwaRF, CH2M HILL, and AWWA (2002) summarize how these elements come together in the *Capital Planning Strategy Manual*. The manual explains that as the mechanism to prioritize capital improvements for new or renewed facilities, the capital improvement program is a shared activity and requires inputs from stakeholder groups.

The manual states:

> Capital budgets are increasing today for most water utilities, driven by the need for new capacity, regulatory pressures, and the need for infrastructure repair and replacement....What is needed to meet current challenges is an organized, documented process, driven by stakeholder priorities and coordinated with other financial and master planning activities....This process must include state-of-the-art analytical tools that allow prioritization of projects based on return on investment and aligned with the utility's vision.

Prioritization, then, is one of seven steps in the capital improvement program process recommended in the manual.

The other steps are defining the mission, framing the problem, establishing the data management protocol, identifying and screening, conducting financial planning and obtaining approval, and implementing the program.

Reasons for capital programs in water utilities are specified in the *Capital Planning Strategy Manual*, and generally include

- Growth—providing water supply to new areas
- Obsolescence—replacing old, undersized pipes
- Condition—renewing areas where pipes have deteriorated
- Regulatory controls—upgrading to meet stricter water supply rules
- Increase in reliability or service—spending to increase security

The manual notes that the consensus at a related workshop (held before the manual was published) was that "utilities are nowhere more different from each other than they are here [America]." Because each distribution system has different characteristics, no two prioritization plans can be the same. Financial and economic objectives that a utility may consider include minimizing life-cycle and financing costs, avoiding major disturbances, allocating costs between parts of the customer base and generations of customers, avoiding premature reinvestment, flattening demographic waves of costs, or sustaining a minimum level of pipe performance.

The investigators found that for capital improvement program prioritization, utilities should use better methods based on stakeholder-driven criteria. Broad types of criteria are

- Technological—Does the plan promise to meet goals effectively?
- Legal—Is the plan legal and does it pass a test of legal feasibility?
- Political—Is the plan politically feasible?
- Institutional—Can the owner implement the plan in the long term?
- Financial—Does the plan have an acceptable rate of return?
- Risk and vulnerability—Does the plan bring unacceptable risk and vulnerability?

- Social—To what extent are people negatively affected?
- Environmental—Does the plan introduce negative environmental consequences?
- Economic—Does it pass a benefit–cost test?

As there was not much literature on capital improvement program development at the time, the AwwaRF, CH2M HILL, and AWWA (2002) investigators conducted a survey. They found that the following criteria were used to prioritize water supply system improvements:

- Provides service to new customers
- Improves service to existing customers and reduces service interruptions
- Supports economic development objectives
- Benefits local economy
- Achieves compliance with regulations
- Reduces or eliminates threats to public health
- Produces revenue
- Reduces operating costs
- Supports other plans
- Brings in outside funding

Of these, pipe renewal might improve service to existing customers, achieve compliance with regulations, reduce operating costs, and protect public health, but it might not compete well in these categories and it would not normally score high in the other categories.

Examples of Planning Processes

During this project, we found a number of excellent examples of how utilities organize their overall planning processes. Planning tasks divide according to organizational function into O&M, finance, engineering, planning, and information management. Eventually, these functions merge into a viable MMS, engineering and planning tasks that supply input to the capital program, and a finance function that takes care of any budget and accounting work required. These should yield answers to the questions raised by all utilities: Which pipes are most likely to fail? What will the consequences be? What are the alternatives for renewal? How should priorities be set?

In the United States, for example, large utilities have undertaken sophisticated organization-wide approaches to distribution capital planning. EBMUD's comprehensive pipe management program includes a new pipeline model based on AwwaRF research. It utilizes a district mapping system and focuses on CIP, which is 40% of its system and has 80% of all leaks. The concept is that pipe performance depends on soil conditions, installation practices, local topography, and so forth. The model assumes that these factors are consistent within small geographic areas, and that the distribution system is divided into 1,400 individual areas. Performance is projected for each area and combined. EBMUD's program goal is to replace pipe when replacement is more cost-effective than continued maintenance. Factors considered include previous leak history, seismic hazards, customer impacts, field staff input, and analysis of the cost-effectiveness of pipeline segments. District representatives use advanced models, such as an online GIS (Clinton 2003).

The New Orleans (La.) Board of Water and Sewerage has implemented a comprehensive system. As explained by Becker (2003), the board's approach, called the Water Master Plan, includes system understanding and characterization, model building and calibration, development of performance criteria, system evaluation, a leakage detection program, alternative analysis solutions, and the capital improvement plan. Performance criteria were developed to evaluate existing distribution system performance, identify facilities in the system that fail to meet the criteria, develop alternative solutions, and determine system improvements. The board uses models such as GIS-based hydraulic software and the KANEW model to quantify the inventory. A prioritization or "micro" model was used to establish structural criteria, establish weighting factors, create the prioritization model and perform prioritization, review and validate the results, and modify the structural criteria and weighting factors based on results. Primary criteria for prioritization were breaks and leaks, age, water quality, soil conditions, importance, and water quality. Secondary criteria were customer impact, location, and material.

WSSC has a process that has been under way for a number of years, as Woodcock (2003) and Habibian (2003) explained at the project workshop.

The South Florida Water Management District has the Southeast's largest water control system, with more than 1,000 mi of canals, 700 mi of levees, hundreds of gates and water control structures, and dozens of pump stations. The district's 2002 estimated capital projects backlog was $387 million, but it budgets only $5–$7 million annually. In 2002 the utility had staff reductions and increased costs because of deferred maintenance, so managers undertook a staff redirection and reclassification effort based on O&M needs. They recognized that capital projects, if not refurbished, will become more expensive as emergency projects. District representatives are seeking to develop life-cycle cost ratios, validate the 50-year planning process, determine future O&M costs, and identify a dedicated funding source. Activities included benchmarking the existing workload, analyzing the workforce utilization, and standardizing maintenance procedures and other activities aimed at optimizing the O&M program. The estimated O&M budget for the next 50 years is being developed using a decision table; evaluation criteria for prioritization; and work from engineering, procurement, and budget.

The City of Portsmouth, Va., has implemented a renewal program for water and sewer lines. At the project workshop, utilities director J. Spacek distributed literature about the renewal program, describing how the city was renewing its 550 mi of water distribution pipeline because the reliability of the system was in question. The program supports the city's mission for economic development, fiscal strength, and neighborhood quality. The timetable began in 1870, when system construction began. In 1988 a utilities report identified replacement needs, in 1996 the city initiated a strategic financial plan, and in 2001 planners conducted an asset valuation study.

Several other countries contributed material for the project (or we were able to study their approaches through the published literature). For example, Sydney Water in Australia has a formal hierarchy of capital plans by infrastructure sector and by area. The utility begins with an inventory of pipe by class and year installed. Next, the plan states that the failure of the pipe is a certainty, but when it will fail is an uncertainty. Planners prepared a risk-consequence matrix showing that the highest risk mains were their older CIP inventory. Management decisions are how much to spend, which pipes to renew, and how to set priorities. Sydney ranks pipes by risk scores according to potential consequences and likelihood of failure. Utility representatives take steps, such as the use of a desktop method to screen pipes likely to fail, to manage the likelihood of failure and the consequences. Sydney Water is now engaged in further research on condition assessment. In addition, the utility is conducting a pilot project that will use GPR, soil sampling, external scanning of CIP, ultrasonics on steel pipe, and coupon sampling. The utility is using

KANEW to forecast financial needs and working more to develop their risk procedure. Finally, a pig is planned for use in inspecting large-diameter mains.

The European Commission conducted a comprehensive research project on pipe management. Seven European countries participated—Germany, France, Italy, Portugal, Norway, the Czech Republic, and England. The research questions examined the condition of a specific pipe and the entire network, the most vulnerable pipes, future failure rates, appropriate rankings for rehabilitation projects, future investments needed, and optimal management of water distribution networks. The CARE-W tool kit includes performance indicators, failure forecasting, reliability forecasting, and rehabilitation planning (Hafskjold 2003).

A number of individual papers refer to specific cases within Europe. For example, Portsmouth Water plc, Havant, U.K., writes about its approach to replacing water mains. This utility sought to replace 1% per year, or 32,000 m. Its point-system prioritization program considered burst frequency, corrosivity of soil, importance of main, age, and material (Neve and Hedges 1998).

ANALYSIS OF PIPE NEEDS BY YEAR

Improving underground assets is not as glamorous as more visible capital improvements. During the capital budgeting process, utilities must determine how the need to renew underground assets competes with alternative investments and convince stakeholders to invest in renewal projects. This requires using techniques such as benefit–cost analysis.

In developing plans, the first question is "What are the needs?" To answer this question, planners must know the current situation and criteria with which to compare that situation. Methods of preparing a needs assessment start with inventory and condition assessment; then the condition is compared with minimum or desirable levels of condition. The needs are the expenditures and actions required to bring condition to the levels needed. This can be done with tabletop work (i.e., preparing a table to compare levels and needs), or with computer-based tools, such as KANEW and Nessie.

KANEW

The KANEW model was designed to provide a method and software for a "predictive distribution system condition assessment model." However, the model is for a system, not individual pipes, so it can be used to quantify costs but not to prioritize repair, rehabilitation, and replacement programs. In a review of the KANEW software, Deb et al. (1998) write that KANEW gives utilities a tool to develop long-range pipe rehabilitation and replacement strategies.

KANEW works by importing data about the network. Data are differentiated according to year of installation or rehabilitation and with respect to aging behavior. The most important criteria are age, material, diameter, and bedding quality. Next, "survival functions"—based on records of past failures and on expert opinion about survival of newer pipes—are determined to predict life expectancy. Taking uncertainty into account, the model offers pessimistic and optimistic scenarios for pipe survival.

KANEW has been tested in North America and Britain. It predicts the quantities of categories of pipe to be rehabilitated or replaced on an annual basis. As a macro model, it does not provide location-specific rehabilitation and replacement information. The criteria for the water mains are age, material, diameter, and bedding quality. The model is user friendly and yields different sets of graphical and tabular outputs.

For this project, we identified characteristics of the North American water distribution systems in terms of rehabilitation and replacement needs. Statistics are available for North American pipes by region, length, average number of miles replaced per year, and average number of miles installed per year for expansion (AWWA 1997, 2002).

The main advantage of KANEW is to allow the utility to use a forecasting system, enabling it to budget over the long term. The most important step in the development of rehabilitation and replacement strategies is the preparation of a comprehensive water main inventory database, preferably incorporated in a GIS framework.

According to Hope (2003), the KANEW model offers a good method for determining a water main replacement rate, based on the age and material of the main. Hope, the vice chair of AWWA's Infrastructure Management Committee, presented tools for developing and implementing a water main asset management plan. He projected the estimated need for replacement along with the average replacement rate at the present. He pointed to the need to balance risk and investment by developing curves for the level of service of assets, along with the cost for maintaining that level. Limitations of the KANEW are that it is very general, not community-specific, and is based on limited variables. To develop a valid model, accurate data must be available. Hope mentioned the lack of pipe characteristic data, at probably only 60%–70% of what is in the ground. Furthermore, the KANEW analysis does not take historical data into account but develops rates based only on estimated averages.

Nessie Curve

In AwwaRF's Nessie study (Cromwell et al. 2001), utilities polled gave an 85-year average life for pipe segments, with many water mains lasting more than 100 years. Much of the present infrastructure in the United States was laid during building booms before and after World Wars I and II. As many of these water mains move into the latter stages of their lives, we can expect "echo booms" of replacement. Australian utilities have graphically represented these replacement booms with what is known as a "Nessie curve." The Nessie curve shows waves of increasing investment needs for utilities. It aims to predict when these waves will occur and how they can be avoided. Creating the model requires a great deal of data on pipe characteristics. The investigators concluded that North American utilities are migrating from reactive to proactive approaches and that using Nessie will give utilities planning data that uses "decay curves" derived from statistical analysis of break data to forecast repair, rehabilitation, and replacement needs for each class of asset. This can then be projected to create the Nessie curve of future investment needs.

Another recent AwwaRF project yielded information on costs of infrastructure failures (Cromwell et al. 2002). The goal of these authors was to study existing cost estimation in other utilities and industries and develop a cost-estimating model for water infrastructure failure. They point out that water main infrastructure failures result in tangible and intangible costs, and that a cost-estimation methodology should consider both social and asset costs. They write

> This report summarizes literature related to estimation of the social costs of service interruptions and when possible service interruption directly resulting from infrastructure failure. The literature has been examined as background and guidance for developing the Grand Central model to better estimate the comprehensive costs of infrastructure failures.

The Cromwell et al. (2002) report gives an overview of cost-estimating approaches for electric power outages, traffic disruption and travel time costs, flood damage, water supply disruption from seismic events, and natural gas service disruption. It includes a disk and a user guide for the Grand Central model.

WHICH PIPES TO RENEW—RATING INDIVIDUAL PIPES

Determining which pipes to renew requires utilities to move from the macro to the micro level. Research shows the prevalence of point systems, such as that in Columbus, Ga., whereby individual pipes are rated by criteria such as age, break history, material, water quality, and location.

Early in the research on individual pipe ratings, a project funded by USEPA and carried out at Massachusetts Institute of Technology yielded several papers about capital budgeting models and pipe prioritization schemes that could focus on single pipes. Karaa, Marks, and Clark (1987) presented a linear programming approach to solving the capital budgeting problem. All approaches like this depend on knowing the return from rehabilitating or replacing individual pipes. This problem was approached by the team and presented in other work, such as Andreou, Marks, and Clark (1987), which modeled break failure patterns down to the individual pipe. To address individual pipes, this team used hazard rates that depend on covariates reflecting various pipe and environmental characteristics and that avoid having to hypothesize relationships between break rates and pipe ages. The team divided incidents into early-stage breaks, which assign failure probability as a function of time, and later-stage breaks, where a different approach is taken that depends on break rates of pipes. This research seems promising but needs further followup if utilities are to use it on a practical basis.

DECISION ANALYSIS AND PRIORITY SETTING

The Decision Analysis Problem

Management science tools have improved decision analysis a great deal, at least in terms of "crunching the numbers." In fact, a whole field of research—decision theory—is aimed at decision analysis, and, although it is not possible to review it all, we can describe how it applies to the prioritization of pipe improvement programs. However, the validity of the methods depends on having valid data or the garbage in–garbage out problem arises. Even if data are valid, they still depend on the right goals, criteria, alternatives, and other steps. Because of the many "ifs" in decision analysis, decision makers tend to rely on intuition instead of number crunching. Having said this, we should still be aware that decision analysis offers powerful tools for problems such as prioritizing investments.

Prioritization Methods

The *Capital Planning Strategy Manual* (AwwaRF, CH2M HILL, and AWWA 2002) describes decision analysis as the most advanced method of setting priorities. However, all methods of setting priorities are "decision analysis," and the manual is referring to using advanced methods which often fall short because of inadequate data. It states that the purpose of prioritization is "to complete an unbiased evaluation and prioritization of capital projects to allocate limited available capital to the projects that will yield the greatest benefit."

The manual presents a framework where the utility selects from three alternative ways to prioritize, ranging from less sophisticated to more sophisticated: voting, spreadsheets, and decision analysis. It is important to remember, however, that more sophisticated will not mean better unless the criteria and data are valid.

In any of the methods, criteria must be established, data about the problems to determine remedial actions must be valid, and alternative solutions must be formulated and evaluated. If these three requirements are met, the choice of method depends on resources that can be expended on it.

Selecting a Decision Method

Voting is recommended for small systems where good information is available; where project needs closely match resources; and where only a few, generally noncompeting, objectives exist. In voting, a protocol that depends on the local situation is set up.

A spreadsheet or matrix can be used for a short list of less than 30 items, with multiple objectives. In a matrix evaluation, the evaluation criteria are selected and weighted, performance measures for the evaluation criteria are developed, and projects are scored and ranked.

Decision analysis is the more sophisticated approach that is used for projects with multiple stakeholders, competing interests, and needing a defensible process.

Decision Analysis

Many variations of decision analysis models are in use, and they all require intensive facilitated stakeholder involvement. The decision analysis method recommended in the *Capital Planning Strategy Manual* (AwwaRF, CH2M HILL, and AWWA 2002) is multiattribute decision analysis, a structured evaluation of the risks and benefits of a decision compared to the costs. The manual quotes a 1986 National Academy of Sciences report that claims this approach states clearly and systematically the assumptions, judgments, preferences, and trade-offs that go into a prioritization decision.

The steps to the multiattribute method are selecting evaluation criteria, weighting evaluation criteria, developing performance measures for evaluation criteria, educating stakeholders on projects, scoring alternatives, evaluating costs and risks associated with projects, ranking alternatives based on a value-to-cost comparison, and remaining wary of pitfalls.

It would be convenient if plans could be reduced to just numbers, but public decisions involve many issues that are not amenable to numerical comparisons. Therefore, the manual acknowledged the practical issue of political trade-offs. The authors write

> Realistically, the existence of political "deal-making" is probably a given in most utility environments. Capital planning decision processes and models are not intended to replace or eliminate the traditional role of politics. Rather, these models can complement the political process; they provide a starting point for negotiations and greater insight into the potential trade-offs between alternatives.

This is sometimes referred to as "trade-off analysis." Once a renewal plan is decided on, continual evaluation and review is necessary. Ongoing evaluation is required to ensure that best practices are being carried out. Ironically, each pipe failure can be useful in developing better predictive models. Each distribution system will be more accurately modeled as information is gathered on expected pipe lifetimes. When action is taken to improve distribution system integrity, the utility should measure the results and make appropriate adjustments.

EXAMPLES OF DECISION ANALYSIS MODELS

The theory of decision analysis was summarized in the preceding paragraphs. During the project we found a number of examples of decision analysis models in use and summarize them here.

O'Day et al. (1986)

The report by O'Day et al. (1986) presented a coherent approach to planning and prioritization for water main improvement. Their system was labeled a "Water Main Decision Planning System." It includes the following: assessing and monitoring system condition, identifying and assessing problem mains, prioritizing mains for repair or replacement, and minimizing public health problems. The authors present an economic model for decisions. They recognized that, although models are being developed, no proven method is available for classifying mains for replacement, and utilities must rely on information they collect, along with subjective judgment, when making replacement decisions. Although more research has been done, this finding still rings true.

The O'Day team describes the reactive and proactive approaches. In a reactive approach, problem mains can be rated according to a point system or benefit–cost ratios. The authors describe how utilities used this approach and present rating criteria and a flowchart. In all cases, the point rating system did not make the final decision, but it provided guidance to a management decision made using judgment.

The proactive or "predictive evaluation approach" allows utilities to also study mains that have not yet developed problems. It uses a water main inventory and database.

This approach, in contrast to the reactive approach, uses data to predict future problems. Of the case studies presented, only one, Philadelphia's, used a predictive approach. Philadelphia relied on a condition assessment model that, as the authors point out, had not yet been fully tested or proved to be effective. Philadelphia's model applied to older gray cast-iron mains and considered pipe, environmental, corrosion, and load conditions.

The authors conclude by advising utilities to consider the economics of using a water main database to create a predictive system. This is one of the recommendations at the end of this report.

Economic Analysis Models

In a general sense, economic analysis is used to compare costs and benefits of alternative courses of action and presents them in forms such as benefit–cost ratio, rate of return, or break-even analysis. A number of studies have presented such models. For example, Walski and Pelluccia published a paper in 1982 with an economic model. More recently, Cromwell et al. (2002) published a report for AwwaRF titled *Cost of Infrastructure Failure,* with data on failures that can be used in economic studies. Loganathan, Park, and Sherali (2002) also presented a threshold break-rate equation that shows the critical break rate for optimal pipe replacement. This method is based on repair and replacement costs and calculates the point at which it becomes more economical to replace a pipe than to repair it. The advantage of this model is that no break rate is embedded in the equation, but only costs and discount rates. Using these data, the model simply generates a rate at which the pipe should be replaced. In an article in *Journal AWWA*, Cabrera, Miguel, and Francisco (1995) discuss "Network Maintenance Through Analysis of the Cost of Water." This article proposes a method based on the history of repairs carried out on a network. The method allows an estimate to be made of the time at which a program for detection and repair of leaks should be started, as well as when the network should be replaced. Sensitivity analysis is used to assess the influence of certain parameters.

A related category of model uses a mathematical technique, such as linear programming or a search method, to find an optimal strategy. The literature contains many such models, but they find their way to the inside of software packages, rather than being basic tools themselves. For example, Dandy and Engelhardt (2001) studied the use of genetic algorithms to schedule water pipe replacement. The study focused mainly on economic criteria, including O&M and optimized capital costs, although the authors note that genetic algorithms could be used for multicriteria problems. They conclude that the genetic algorithm was able to identify appropriate pipes that needed the most attention. This is an example of optimization, but again, it depends on valid data for individual pipes.

Capital Allocation and Financial Models

Capital allocation and financial models are similar to economic models. They are aimed at deciding how to spend fixed or variable budget amounts, rather than finding the best rate of return, regardless of the amount of money involved. Rate of return investment models that consider risk and investment alternatives fit into this category.

Washington Suburban Sanitary Commission's Approach

WSSC's approach shows a comprehensive, overall way to address main replacement. Information on WSSC's program was furnished initially during our March 2002 workshop, which WSSC hosted (Woodcock 2002a). Tucker and Corriveau (2003) described WSSC's overall program.

Beginning in fiscal year 2002, WSSC initiated a program to replace water mains in a two-county region. The plan was to replace 8 mi/yr, a 30% increase, but still only about 0.1% of the overall system. By 2007, this will be ramped up to 20 mi/yr. The decisions required WSSC to use its "Legacy Maintenance Management Information System" and data and GIS more aggressively. The utility's consultant developed a MapObjects (ESRI, Redlands, Calif.) application for WSSC's Systems Infrastructure Group, which enabled it to use its historic and spatial data for capital planning.

The Systems Infrastructure Group resides within the Engineering and Construction Team at WSSC, with responsibility for investigation, analysis, planning, and design, along with contracts for water main replacement, water main cleaning and lining, and sewer rehabilitation. To achieve these, the group manages design and construction standards and guidelines and provides technical support to other WSSC groups.

Earlier, the water main replacement program utilized manual methods where the Systems Infrastructure Group identified high-maintenance pipes on an area-wide and localized basis. The group decided to use information technology to a greater extent and developed a Utility Information Management Decision Support System (UIMDSS). This MapObjects tool uses a GIS and maintenance history data to feed a replacement and rehabilitation algorithm the group developed to pinpoint candidate pipes for immediate or long-term replacement or cleaning and lining.

The replacement and rehabilitation analysis is initiated from the main interface of the UIMDSS. The GIS tool enables the analyst to select the space scale for the work.

Maintenance activities are specified, including discolored water, performance, broken main repairs, fire hydrant leak repairs, main leak repairs, service leak repairs, valve leak repairs, and routine flushing.

Next, the UIMDSS analyzes the selected pipes and displays the results as immediate replacement candidates, long-term replacement candidates, cleaning and lining candidates, or pipes not meeting any repair and replacement criteria. The maintenance data, dating from 1974, are from WSSC's mainframe MMIS. There are about 10,000 mi of pipes and 425,000 customers with records documenting maintenance activities, including corrective maintenance, complaints, preventive maintenance, and rehabilitation.

The replacement and/or rehabilitation algorithm applies a weighting factor according to type of maintenance and generates a score for each pipe. According to WSSC, "engineering principles and statistical analyses" define criteria and thresholds for each category. Water mains defined as replacement candidates had greater than or equal to four main breaks in the last 5 years, or at least six main breaks since the pipe was constructed, or an overall score of 20 (considering overall maintenance impact). Long-term replacement candidates had between two and four main breaks in the last 5 years, or from four to six breaks since construction, or an overall score of 12.

Cleaning and lining candidates are CIP, those that have not been cleaned and lined, those constructed between 1935 and 1967, those with less than three breaks, and those having an overall score of 2.25 or less. No-action pipes are all others.

In compiling the overall scores, WSSC gave weights to maintenance activities, given in Table 6.1. The overall score is the sum of maintenance counts, multiplied by the weight factor, divided by a unit pipe length ratio, calculated by dividing the pipe length by 1,000 ft.

Table 6.1
Weights given to maintenance activities at WSSC

Maintenance activity	Factor
Repair broken main (within the last 5 years)	1.0
Repair main leak	0.85
Repair broken main (within 2 years of cleaning/lining)	0.8
Repair broken main (not in the last 5 years)	0.75
Repair valve leak	0.5
Routine flushing	0.25
Repair service leak	0.25
Repair fire hydrant leak	0.25
Discolored water	0.25
Performance	0.25

Courtesy of Los Angeles Department of Water and Power

Figure 6.3 Sliplining project

LADWP's Approach

Mavrakis (2003) presented LADWP's methodology at the project workshop. Figure 6.3 shows one of LADWP's projects. The utility uses an A–F rating scale with age-related deterioration scores. For pipelines, the maximum service life was assumed to vary from 70 to 210 years, depending on pipe material and soil corrosion index. Three soil types are considered: severe corrosion (30%), moderate corrosion (50%), and mild corrosion (20%). Pipeline grades were determined separately for steel and CIP/DIP. Trunk lines were evaluated, and their lengths were taken from the trunk line hydraulic model. Some 743 mi of trunk line (16 in. and larger) were evaluated on an overall basis at "B," with a unit cost of $13/in. of diameter per foot. Of the lines, some 70 mi were rated D, F, or critical. The present value of replacement cost of all trunk lines was $1.67 billion. This analysis yielded an estimate of $54.5 million needed now for replacement of critical mains and $10.1 million needed annually to keep up with critical replacements over the next 20 years. Considering all assets, including tanks, reservoirs, treatment plants, and trunk lines, a cash flow replacement schedule similar to the Nessie curve was produced. Total average replacement needs are $16.2 million per year.

Columbus, Ga.

Columbus has a large inventory of small CIP and galvanized steel pipe (3 in. and smaller) that needs renewal. The inventory is estimated at 167 mi, with a replacement cost of $52.4 million. Columbus Water Works undertook a GIS-based study to identify lines for replacement and used several information systems to obtain information on breaks and low-pressure and water quality complaints, to go along with age and other data that the utility uses in its prioritization scheme. The scheme is based on a voting system. Priorities, in order, are customer complaints, water quality, maintenance, fire protection, materials, age, and pressure. Utility managers also decided to replace some of the service connections as they replaced small lines. The winning bid was $58.67/lf in year-2000 dollars. They also compared their capital improvement program to Nessie, which requires assignment of service lives to categories of pipe. Age data were compiled from different sources, including employee interviews, but the data varied and indicated some uncertainty. Lessons were that the GIS is a useful tool for this kind of study and for asset management, age dating is useful, database access is required, and just getting the pipe data represents good progress.

Edmonton, Alta.

The City of Edmonton began a reactive prioritization program in 1985 (Seargeant 2003). In 1985, Edmonton experienced approximately 1,600 main breaks. Through a program in which a surcharge dedicated solely to replacing main breaks was added to customer bills, the number of breaks was reduced to 468 in 2000. The surcharge equated to about $3 to $4 per customer. Seargeant mentions that Edmonton applies a proactive program, which used hydroscope results, a hydraulic model, and coordination with other utilities, in some cases. However, the noted success was seen almost entirely from a reactive program. Replacements were based on service level in which all pipes with a break frequency of greater than 5 breaks/km/yr, based on a 5-year average, were replaced.

Comprehensive Software Tools

Comprehensive software tools contain approaches to decision analysis. For example, MapObjects is used within CH2M HILL's infrastructure capital asset management system. The system combines a GIS with asset management in cataloguing asset condition within a capital program. Booth and Rogers (2001) explained that the model predicts capital expenses for the life of each asset, evaluates funding strategies, and displays expected condition and annual costs based on scenarios.

Some software packages bundle applications across public works categories. For example, CASS WORKS of RJN Group (Wheaton, Ill.) was used to centralize data for the Everett, Wash., Public Works Department. Eight modules—warehouse, plant, water, sewer, storm, pavement, physical inspection, and image—were implemented through a central relational database that accumulates asset history. The Everett Public Works Department adapted the software to its own business practices, rather than change practices to meet the software configuration.

Others report similar approaches. Hope (2003) developed an approach for prioritizing water main replacements that assigns a priority action number (PAN) to each main based on agreed service-level goals. Several service-level criteria were established including pipe age, leak history, water quality, criticality, hydraulic performance, corrosion, material, system pressure, and location. Once PANs are determined for each category, categories are weighted according to importance. After the numbers are assigned, a prioritization model called OWASA is created. In it, a variety of data sets—such as operational work orders, a sewer system hydrologic model, customer service billing information, and a water system hydraulic model—are integrated into one geodatabase. From this integrated database, a water main prioritization model can be developed, along with an overall utility capital improvement plan. With all of these data integrated, sections of pipe are designated as highest priority for replacement.

Hope (2003) presents procedures for prioritizing by using data collected during water main breaks. Based on a literature search and a survey (conducted in North America [37 replies of

70 surveys] and Europe [28 replies of 44 surveys]), Hope grouped methods into four categories: deterioration point assignment methods, break-even analyses, failure probability and regression methods, and mechanistic models to simulate the deterioration of a pipe over time and the loads to which a pipe is subjected. Types of data to be collected are described, and the Hope team developed and tested a mechanistic model for CIP to prioritize pipes. The model is enclosed on a compact disk with the report. A data flowchart and a sample collection form are included, and statistics and ideas for developing a database with information from water main repairs are reported. In the report on a decision support system, the Hope team considered criteria for selecting technologies and pipe materials and developed a decision support system to select the most appropriate renewal technologies for a distribution system. Twenty utilities were surveyed and information on main renewal techniques and factors to be considered is presented.

The three parts of the mechanistic model are identifying factors, prioritizing, and selecting the appropriate technology for renewing the pipe based on the pipe characteristics and site. Unit cost data are required.

Research by Canada's National Research Council has produced a number of papers and reports about prioritizing distribution system renewal. The council is developing a system called WARP (Water Mains Renewal Planner). This model, and the accompanying research, has several useful features. First, the researchers recognize that by studying water main failure patterns over time, insight into the causes and solutions of failures can be gained. For this, they rely on statistical techniques that were described in a *Journal AWWA* article by Kleiner and Balvant (1999). The WARP models time-dependent deterioration factors such as aging and environmental factors to predict break rates, and introduces a method for considering cathodic protection strategies along with the time-dependent aging and environmental factors. The model yields predictions of needed investment patterns for system-wide application. The investigators are also studying ways to add modules to plan for individual pipe replacements. The team is part of a larger research effort mounted by National Research Council Canada's Institute for Research in Construction. Its Web page contains a number of papers that can be downloaded (National Research Council Canada, Institute for Research in Construction 2002).

Expert Systems

Expert systems are a form of software that prioritizes investments. Hahn et al. (2002) did a knowledge base compilation by interviews with experts and developed an application for sewers. The model considers failure mechanisms and defect parameters, including installation history, material, age, soil type, groundwater, loads, exfiltration, inspection history, corrosion and erosion parameters, and so forth.

METHODS AND SCHEDULE FOR RENEWAL

Methods for renewal comprise another management decision, and these methods are discussed in the next chapter. The schedule for renewal is an output of priority setting, when compared to available funds in the capital budget. Utilities also must select funding sources.

PRESENTATION TO DECISION MAKERS AND THE PUBLIC

Regardless of the method used for prioritization, the presentation format is an important ingredient for success. We did not evaluate these but noted different approaches among utilities.

In addition to management decision making, public involvement is also important. For example, education and outreach enabled San Diego's water department to overcome opposition and win public support and council approval of its $773-million plan to repair, replace, and upgrade its water treatment, storage, and delivery system (Steirer, Katz, and Griffin 1998).

RESEARCH AND DATABASE MANAGEMENT

The other part of the decision tree is research and database management, which were combined to emphasize that research must rely on the main break and pipe condition database.

Staffing refers to how the research is done by experts within the utility. The scope of this project did not include recommending how to organize and staff and utility research effort for distribution capital renewal. However, we found that utilities are demonstrating success in assigning the relevant tasks to planners, engineers, database specialists, and others, and in finding ways to integrate their findings into plans.

Spatial analysis is a critical component of a utility research/database program, and two basic choices are available: mapping or a GIS (which is automated mapping). We did not review software platforms for GIS, but they are evolving rapidly.

Database management refers to how the data are organized and used. The two basic strategies, paper records and software, are obvious, but what is not obvious is the discretion utilities have about developing their own databases or using vendors to supply them. The project did not extend analyzing ways to collect and manage information for pipe renewal, but we observed that utilities are making progress in learning to collect, organize, and manage data. We believe that this constitutes an area for further research that would identify best practices and methods for planning the data program, including identifying different kinds of data, collecting the data, organizing and analyzing data sets, and making data available to different parts of the organization.

As an example of research in data management, Kleiner and Balvant (1999) published an article in *Journal AWWA* on "Using Limited Data to Assess Future Needs." Their paper discusses how data on pipe classes can be partitioned to predict failures. Their approach showed how far we have to go before data can actually be used to prioritize on a proactive basis. They point out that much more data is required before a single pipe analytical approach can be used, such as that proposed by Andreou, Marks, and Clark (1987).

Decision support system refers to a packaged approach to analyzing information and making recommendations. Deb et al.'s (2002b) work on decision support systems was summarized earlier in this chapter.

Workplan and strategy should lead to the utility's problem-solving strategy. The general process is to identify the problem, create alternative solutions, evaluate them, decide on actions, and implement actions.

WORKING GROUP REPORTS

Two working groups at our project workshop focused on planning and prioritization. They noted the options available for repair, rehabilitation, and replacement; trenched and trenchless work; and planned, unplanned, or opportunity work. "Opportunity-based" work is renewal that takes place when breaks and leaks are discovered. Another common form of this type of work is renewing pipes when roadwork is scheduled, thereby avoiding two pavement cuts.

The groups confirmed that programs for planning and prioritization are important to enable utilities to proceed with renewal programs. To develop comprehensive approaches to planning and prioritization, many elements must be considered. The main issues are criteria for prioritization, critical versus noncritical links, the right level of renewal, performance standards, and spending amounts (including whether constituents can afford it, without cash flow surprises). The importance is linked to system reliability, adequacy, efficiency, cost-effectiveness, avoidance of financial crisis, and customer service.

Obstacles include collection of relevant data, limited training, limited funding, limited staffing/understanding, and public opposition (to rate hikes and such). Options for overcoming obstacles are clearly informing the public, training staff, developing performance standards, improving reporting and record keeping, making a clear case to decision makers, preparing an asset management plan with continuous monitoring, and integrating.

Effective planning requires a good water distribution model, a plan, a good decision support model, life-cycle analysis, a GIS, system needs prioritization, critical element identification, coordination with other agencies, maintenance plans, feedback from field staff, and a cash flow plan.

All of this must be connected with the construction program. The utility should foster relationships with suppliers and must offer financial incentives. Other points made by the working groups are the need for a system inventory and the capability of downloading information for the operators.

To summarize, the project workshop participants stressed the need for data and that collecting and sharing information should become a way of business. Failure mechanisms with different types of pipe must be known, and best practices (as described previously) should be employed.

CONCLUSIONS

This chapter shows that we know much about the processes of planning and budgeting and about capital strategy. Many analytical methods are available to forecast and analyze needs, if valid data are available. However, we do not know enough about failure mechanics to predict failure and thus do valid economic studies.

The financial risk of water utilities depends on uncertainty with future pipe replacement needs and compliance with regulations. The research on planning and main break prediction show that analytical approaches are available but valid data are missing, which leads to the need for a research project with two goals. First, the research about data needs and attempts to collect and make sense of the data must be evaluated (essentially, this would extend the work of Kleiner and Balvant [1999]). Second, a better approach for data management by a utility should be developed. There would be two levels—one would be utility-wide, a reasonable approach to financial planning, and the other would be a workable way to use data to predict when individual pipes need repair, rehabilitation, and replacement.

CHAPTER 7
REPAIR, REHABILITATION, AND REPLACEMENT

CHAPTER OVERVIEW AND PURPOSE

In this chapter we review the technologies used in repair, rehabilitation, and replacement. These have been reviewed in other publications, and Deb et al. (2002b) summarizes the technologies for decision purposes. Generally, repair, rehabilitation, and replacement practices vary greatly among utilities. All utilities repair broken pipes and leaks, but some are more aggressive than others. Some utilities rehabilitate pipes regularly, and others focus on replacement without much rehabilitation activity. Renewal in water distribution systems is more complex than in wastewater because of the large amount of hardware—such as valves, service connections, fittings, and so forth—involved in water distribution. However, new and innovative methods are coming online all the time.

Maintaining water service while repair, rehabilitation, and replacement are proceeding is a big issue, as is contamination of water supplies (Ellison 2001). Service connections have to be reconnected manually, but the phrase "if you need to dig at all, why not dig it all" may not apply because of the disruption caused by longitudinal cuts. If a method could be found to remotely connect water services, trenchless technologies would have a great deal more potential for water supply, as they do for wastewater.

Topics covered in this chapter are

- Definitions and classifications
- Task map
- Comparisons of technologies
- Maintenance to support repair and rehabilitation
- Repair technologies
- Rehabilitation and replacement technologies
- Protection with cathodic retrofit
- Renewal technologies working group
- Decision tree

DEFINITIONS AND CLASSIFICATIONS

The lines between repair, rehabilitation, and replacement can be blurred, and differences between the words can be significant in preparing needs assessments and capital budgets. In addition, maintenance can include aspects of repair. Grigg (2002) reviewed technologies for maintenance management and explains in that work that when maintenance involves minor repairs, it should be funded from operating funds, but when repairs become major, they belong in capital programs rather than in operating programs. This chapter deals with both minor and major repairs.

Definitions given by Heavens and Gumbel (2002) are presented in Table 7.1, along with suggested modifications, based on our research.

Note the progression from repair to replacement. In repair, functionality is restored; in rehabilitation, a component is restored to its original condition; and in replacement, a substitute

Table 7.1
Definitions of renewal activities

	Heavens and Gumbel 2002	Proposed definitions
Repair	Rectification of local damage	To restore functionality of existing pipe segment(s) after damage or breakage
Rehabilitation	All measures for restoring or upgrading the performance of an existing pipeline system	To restore or upgrade existing pipe segment(s) through extensive work
Replacement	Rehabilitation of an existing pipeline system by installing a new pipeline system, without incorporating the original fabric	To provide a substitute for a distribution pipe or pipe segment

is installed for the original item. The terms "renewal," "refurbishment," or "renovation" are sometimes used as well. In this report, we use *renewal* to include repair, rehabilitation, or replacement. Even with these definitions, the lines can still be blurred, especially those between rehabilitation and replacement.

We know that average replacement rates hover around 0.5%, or once in 200 years, but we lack data on rehabilitation rates. In addition, that data would require a fixed classification system if it were collected. Moreover, we do not know how long rehabilitation technologies, such as CML, will last. We do know, however, that the level of renewal using nontraditional technologies, such as pipe bursting, sliplining, and other trenchless methods, remains low compared to traditional construction methods using trenched replacement.

TASK MAP

The renewal task map in Figure 7.1 classifies work by processes and shows generic and special tasks. These processes can be used to discuss jobs, tasks, and learning requirements, and they show how the workforce can be cross-trained to carry out similar tasks involving different processes and technologies.

The task map is intended to illustrate the generic processes, so that we can compare the levels of technology. Note that all items on the list are not always required; the list is meant to describe work processes:

- Locate and identify the problem: In this beginning phase of the process, the basic decision of what to do is made.
- Plan, design, and contract: In this phase, what to do and how to do it are determined.
- Access conduit space: Work here depends on whether it is a repair, rehabilitation, or replacement, and whether the technology is digging, trenching, or trenchless.
- Prepare conduit space: Here preparation is made for the new or rehabilitated conduit; for example, pipe bursting might open the way for a new pipe to be pulled, or cleaning might be required for rehabilitation.

Dug Repair	Leak/ Break Location		Dig	Clean and Prepare	Repair				
Trenched New or Replaced			Trench or Dig	Bedding					
Process	Locate and Define	Plan and Design	Access Conduit	Prepare Space	Install Conduit	Prepare Joints	Test	Assemble Appur-tenances	Backfill and Disinfect
Trenchless Technology Replace			Bursting Micro-T		Pull Pipe or Lining				
Trenchless Technology Rehabilitation			Trenchless Technology Access	Clean	Reline				
Trenchless Technology Repair			Trenchless Technology Access	Prepare to Seal	Seal				

Figure 7.1 Renewal task map showing special processes

- Maintain services: Services are a critical part of any repair, rehabilitation, and replacement job, and a plan to maintain them is required.
- Provide pipe: The optimum pipe material must be supplied and inspected.
- Pull or install conduit, or line pipe: Here, new or renovated conduit is installed or created.
- Prepare joints: Joints must be restored to full functionality.
- Perform repairs: Here, new conduit may not be involved, but repairs must be performed.
- Test pipe: Repaired, rehabilitated, or replaced pipe is tested to ensure proper operation.
- Assemble appurtenances: Appurtenances such as pumps and valves are reassembled.
- Backfill and disinfect: This is the final phase.

COMPARISONS OF TECHNOLOGIES

Several of the reports or texts we reviewed included tables meant to compare the technologies. Because no standard definitions exist, the content of the discussions varies.

Ellison (2001) presents a table comparing advantages and disadvantages of pipeline renewal methods. His list is comprehensive and includes conventional replacement, CML, epoxy lining, spray-on PE lining, cathodic retrofit, testing plus spot repair/replacement, HDPE sliplining, tight-fit HDPE sliplining, pipe bursting, polyester-reinforced polyethylene, cured-in-place pipe (CIPP) lining, reinforced CIPP, and joint seals. Alternatives are presented somewhat differently by Selvakumar, Clark, and Sivaganesan (2002), who offer a table with similar information, giving size ranges, common materials, costs, and references for the costs. Smith et al. (2000) and Heavens and Gumbel (2002) also presented several figures and tables comparing methods.

MAINTENANCE TO SUPPORT REPAIR AND REHABILITATION

Maintenance systems can include minor repair. See texts by Jordan (2000) and Grigg (2002) for maintenance guidelines. The framework to organize maintenance is the MMS, which takes a systems approach to maintenance. The general functions in MMS are inventory, condition assessment, preventive maintenance, and corrective maintenance. In facilities management, the focus is on computerized MMS, or CMMS; and MMIS is another term. According to Thomas (2001), writing in the *Facility Design and Management Handbook*, "The effective implementation of a CMMS or work management program has become mandatory for facility departments."

In a recent AwwaRF report, Deb et al. (2000) presented guidance for maintenance management. These authors reviewed the literature and surveyed utilities. One conclusion was that maintenance procedures emphasize flow and pressure but neglect reliability and water quality. They point out that water quality can be addressed by maintaining a chlorine residual, structural integrity of pipes, and the cleanliness of the system, including prevention of tuberculation, corrosion, and sediment accumulation. Preventive maintenance, workforce, construction, O&M, inventories and information systems, better design, and excellent capital improvement programs would address reliability issues. They recommend measures for enhancing reliability and water quality, improving O&M, enhancing performance, and managing information.

The report by Deb et al. (2000) is a comprehensive overview. In addition, case studies are presented in the literature and at conferences. At the AWWA 2002 Infrastructure Conference, a speaker from EBMUD explained RCM. An interesting feature of RCM is to move away from proactive maintenance to a "run-to-failure" mentality for some equipment. This, which may lower costs, places a focus on critical equipment and components, which are defined by water quality violations, productivity, regulations, and safety. Higher maintenance cost is a big factor in rates. RCM is data intensive and requires a multidisciplinary team, including maintenance, mechanical, electrical, and instrument engineers and operational personnel (Goodson 2002).

Water quality is not the focus of this report, but AwwaRF has funded recent reports on system O&M to improve water quality. Two examples are *Pathogen Intrusion Into the Distribution System* (Kirmeyer et al. 2001) and *Practices to Prevent Microbiological Contamination of Water Mains* (Pierson, Martel, and Hill 2001).

REPAIR TECHNOLOGIES

The process of repairing a leak or break requires locating the problem, getting organized, and completing the repair tasks. New technologies and methods are emerging for locating problems, in using better equipment for excavation, and in applying materials to repair the problem. Because once a break has occurred, another is likely to occur, so it may be best to replace the broken section. Repair, then, may be thought of as unplanned partial replacement.

Technologies for locating leaks and breaks are described in chapter 5. If a break is repaired with a repair sleeve, as explained in Von Huben's handbook (1999b), the procedure is to locate the break or leak, notify customers, call the local "dig safe" authority (or equivalent), shut off valves, excavate, and install a repair sleeve. Special precautions are taken to disinfect the sleeve and to protect the pipe from contamination. Spot repairs may only use clamps or patches (Ellison 2001).

Internal joint seals can be used to eliminate leaks in joints for pipes of 16 in. and larger. They can also be used to bridge over cracks and similar problems in pipe segments or old connections. A new but expensive method for large-diameter pipe is the use of carbon fibers

Figure 7.2 Joint seal

(Ellison 2001). Linings for PCCP repairs might include adding prestressed tendons and shotcreting and sliplining with welded steel pipe.

The research literature includes works that describe specific techniques and cases. For example, Topf (1990) describes a special internal sealing system for joint repair. The Fort Worth Water Department had problems with joint leaks in a 60-in. reinforced concrete conduit that had been installed in 1926. After repair in April 1987, there were no leaks in the conduit (Bradley and Crumb 1987). Vendors at water industry trade shows regularly display joint-sealing technologies (see Figure 7.2).

REHABILITATION AND REPLACEMENT TECHNOLOGIES

Introduction

In our definition of rehabilitation, cleaning and lining are the principal methods. Sliplining could also be classified as replacement, as some authors do. AWWA (2001) published Manual M28 on rehabilitation of mains, but it also explains repair and replacement work, and covers performance criteria, cleaning, lining, joint seals, and pipe bursting. Following the definition at the beginning of this chapter, rehabilitation is "to restore or upgrade existing pipe segment(s) through extensive work."

Methods for replacement have advanced rapidly in the last decade. They include open-cut replacement—conventional trench and narrow trench—and trenchless replacement using microtunneling, pipe bursting, or sliplining (Smith et al. 2000, Ellison 2001).

In a report titled *Assessment of Existing and Developing Water Main Rehabilitation Practices*, Deb et al. (1990) cover methodologies for evaluating water main rehabilitation as they were understood at that time. Another report followed 9 years later with Deb, Hasit, and Norris

Table 7.2
Technologies for pipe cleaning

Nonaggressive cleaning by flushing	Up to 6 in.
Nonaggressive cleaning by foam swabbing	Very small to over 36 in.
Nonaggressive cleaning by air scouring	3 to 8 in.
Aggressive cleaning by high-pressure water	3 to 8 in.
Aggressive cleaning by pressure scraping	18 in. to less than 36 in.
Aggressive cleaning by abrasive pigging	3 in. to less than 36 in.

Adapted from Smith et al. 2000

(1999) writing *Demonstration of Innovative Water Main Renewal Techniques*. It appears that the term *renewal* is growing in favor, perhaps to replace *rehabilitation*. In the 1999 report, the authors present the latest information on trenchless rehabilitation and replacement, pipe bursting, and pipe lining. Their findings included recommendations on appropriate technologies, which can be presented in the form of a decision tree. They also present details of problems and discoveries encountered during the demonstration project. Recommendations are in the form of guidance in a table of water main problems, consequences, technologies, and suitable improvements.

Pipe Cleaning Table

Smith et al. (2000) summarize technologies for cleaning as shown in Table 7.2. Other writers present similar information. Cleaning pipe for the purpose of improving flow has benefits, but utility managers are not always sure that it is justified. Water quality rules may affect this more than concerns about pressure and flow. Flushing is the most common method to clean pipes (see Figure 7.3), and unidirectional flushing is a newer method. In order of aggressiveness, cleaning methods are soft swab pigging, chemical cleaning, air scouring, abrasive swab pigging, high-pressure water jetting, power boring, and drag scraping (see Figure 7.4). Aggressive cleaning to remove tuberculation may create new problems, especially for pipes that are not subsequently lined (Ellison 2001).

Boyle Engineering Company (2000) prepared an AwwaRF report on *Investigation of Pipe-Cleaning Techniques*. We did not review this report for this project, but according to AwwaRF's Web page, it covers existing pipe-cleaning and relining techniques and discusses the effectiveness of these techniques on water quality improvement, infrastructure reliability, risk management strategies, customer service, and maintenance of reasonable friction factors. It was written to serve as a reference guide to assist in selecting pipe-cleaning methods.

Liners

Pipe can be rehabilitated with a nonstructural or semistructural liner, or essentially by replacing a pipe with a new structural liner or pipe bursting, followed by insertion of new pipe. The literature contains a number of summary tables showing types of liners that include nonstructural lining (cement mortar or epoxy), semistructural lining (close-fit pipe, woven hose,

Courtesy of Sioux Falls, S.D., Public Works Department

Figure 7.3 Hydrant flushing

Courtesy of City of Sioux Falls, S.D., Public Works Department

Figure 7.4 Scraper cleaning tool

Courtesy of Sioux Falls, S.D., Public Works Department

Figure 7.5 Lining machine setup

CIPP, or spirally wound pipe), and structural lining (continuous pipe or segmented pipe). Also, a table of pipe structural characteristics taken from Gummersheimer and Heavens (1998) is presented in Smith et al. (2000). In it, Class I is nonstructural (NS), Classes II and III are semistructural (SS), and Class IV is fully structural (FS).

Ellison (2001) explained the use of two proprietary structural liners mainly for wastewater and based on reinforced cured-in-place technology. One uses a resin-impregnated fabric tube, which is cured in place with steam. Polyester-reinforced polyethylene is like a fire hose and is inserted into the host pipe, then expanded with steam. The host pipe is needed as a structural casing. This is economical up to 8 in., according to Ellison. Figures 7.5, 7.6, and 7.7 illustrate three scenarios for pipe lining.

Cement–Mortar Lining

CML is a proven technology that has been practiced for almost 50 years. Participants at the project workshop gave comprehensive reports on experience with CML. Topics covered were methods, contracting experiences, and lessons learned (Russell 2003, Van Dyke 2003). The experience base with CML is extensive, and utilities are able to access substantial information about it, including AWWA manuals. CML is not without its critics, but the bulk of the experience seems to recommend it as a rehabilitation technique.

Methods of cleaning before CML work begins include hydraulic flushing, use of a steel blade, drag cleaning, and use of a mechanical scraper. Winches pull the lining machine through the pipe and a drag trowel provides a finish. Service laterals less than 2 in. are blown free of mortar before the lining is set. With CML, services are disconnected from the meter and connected to a hose. After scraping and again after lining, the pipe is blown out to get rid of waste. For large-diameter pipelines, mortar is applied centrifugally with a speed buggy and rotating

Courtesy of Louisville Water Company

Figure 7.6 CML project

Figure 7.7 Folded liner

trowels. Chlorinating, testing, reconnection of services, removal of temporary water, and paving access locations are required (B. Britton, personal communication, May 1, 2002).

CML results in a highly alkaline environment next to the metal, virtually eliminating internal corrosion and tubercle formation. Although the ability to do CML has existed since the 1940s, only a few companies do it. Cost for CML ranges from 25%–50% of the replacement cost. CML extends life and reduces leak rates, but we do not know precisely how long CML lasts (Ellison 2001).

Warren, Mason, and Conroy (1990) wrote *In Situ Cement Mortar Lining—Operational Guidelines and Code of Practice*, which is a source document for the *Water Mains Rehabilitation Manual*. The Warren, Mason, and Conroy work discusses remedial action to prepare the system for renovation, the type of cement mortar to use, and the CML process with quality assurance. It also explains cleaning and lining with schematics and presents a technical explanation on how to do the work.

Epoxy Lining

Epoxy lining can be used where soft water will deteriorate CML. It is more favorable for small pipes, such as those found in Britain, but less tolerant of small defects than CML, and it might not stop corrosion. A variation is polyurethane lining, which cures fast and might eliminate the need for bypass piping (Conroy, Hughes, and Wilson 1995; Ellison 2001).

The goal of Conroy, Hughes, and Wilson (1995) was to test and report on an epoxy lining used at a U.S. utility. The group reviewed literature on epoxy lining, including U.K. results, and tested a main in the United States using closed-circuit television and 30-day laboratory testing. The testing showed that leaching of the epoxy in contact with water was well within water quality regulations. The report recommends regulations to allow epoxy lining to compete with CML.

AwwaRF is currently pursuing new projects on the evaluation of epoxy lining.

Sewer Work

Trenchless technologies are often used for sewer work. For example, a sewer job in Knoxville, Tenn., through its historic district, was reported to give minimum disruption (*Public Works* 2001). Another difficult replacement sewer job to replace 50-year-old sewers used microtunneling, jet grouting, and shaft construction techniques in a congested area of Honolulu, Hawaii (Kwong 2001).

Trenchless Technologies

We did not review the status of microtunneling, but it is an important technology in the field of trenchless technology.

Pipe bursting can actually increase diameters (see Figure 7.8). CIP is readily burst, and new applications for DIP and steel are being developed. Pipe bursting is said to have much potential and some risk (Ellison 2001). The literature contains a number of case studies of pipe-bursting projects, especially for sewer work. It is also used for water supply, but the reconnection of service connections remains a problem to be overcome.

The last decade has seen an explosion of new lining and sliplining products. Many, however, were developed for wastewater and cannot be used for pressure pipe (Ellison 2001).

Figure 7.8 Pipe bursting demonstration

HDPE sliplining can be used inside old pipes and is better if there are few lateral connections. HDPE is good for flexibility and fused joints but bad for loss of diameter. It can have thick walls that result in a high pressure rating. Steel plates can also be jacked inside of large-diameter pipes, then welded and cement lined. The annulus must also be grouted. HDPE can also be used in tight-fit applications. All these are newer applications and have not been tested much. The pipe can be put through a die or series of rollers to temporarily reduce diameter or rolled into a "U" shape for insertion. Thinner walled pipe is required for this application, and new tools and devices for it are also required (Ellison 2001).

We found numerous articles and conference papers citing case studies of liners for both water and sewer, and these can be readily located in trade magazines and other literature.

Deb, Hasit, and Norris (1999) discussed three pipe-bursting methods (hydraulic, pneumatic, and static pull) and two lining methods (rolldown sliplining and CIPP with Paltem [Pipeline Automatic Lining sysTEM, Ashimori Co., Ltd., Thailand]) to replace water mains. These authors document the applications by videotape. Now, ductile iron can be pulled through tunnels drilled by horizontal directional drilling.

A new book by Dietrich Stein is scheduled to be printed in English soon: *Trenchless Technology for Installation of Cables and Pipelines* (reviewed by Sterling 2003). Kirmeyer et al. (2000) presented a report on lead pipe replacement.

PROTECTION WITH CATHODIC RETROFIT

Corrosion protection methods open a large area of research, and we were able to summarize only a few studies (see chapter 5).

Cathodic protection retrofits offer a promising avenue to extend the life of metallic pipe. We found a variety of reports and opinions in our workshops and in the literature, and, as discussed in chapter 5, discussions do not always distinguish well between types of corrosion. Some believe that most water pipe will last indefinitely with cathodic protection, and Ellison (2001) reports that cost of the new pipe is less than 1%.

One research issue is the effect of electrical grounding on pipe integrity and shock hazard (Duranceau, Schiff, and Bell 1996). When pipe is used in the grounding system of buildings, alternating current and direct current have effects on corrosion, raise possible water quality concerns, and bring safety for utility workers into question. The discussion mostly centers on shock, with some indication that grounding current increases corrosion.

RENEWAL TECHNOLOGIES WORKING GROUP

At the project workshop, the working group on renewal technologies observed that North American water utilities are reluctant to use available technologies. Reasons are lack of awareness, cost-effectiveness, liability, and uncertainty about sustainability.

Some large utilities use DIP and are not eager to change to materials suitable for trenchless technologies. Options mentioned were reducing the cost by assembling larger projects; providing training to utility personnel and education for management, politicians, and customers about the benefits of new technologies; and using the private sector for construction of renewals.

Research suggested included case histories, technology evaluations, analysis of reluctance to adopt practices, and review of European practices. The case histories would increase awareness and confidence of water utility managers for using new pipeline renewal technologies and promote using new technologies. The studies would evaluate installation and performance of renewal technologies and identify critical factors that lead to successful projects.

Evaluating emerging technologies for water main renewals would address the lack of knowledge of technologies, identify advantages and constraints, and evaluate installation and performance.

Another project would evaluate barriers to using new technologies in North America and develop potential solutions and associated methodologies. At the workshop, a project was also suggested for the evaluation of long-term effectiveness and performance of field-applied CML.

DECISION TREE

A utility confronted with the need to renew water mains has a number of decisions to make. Deb et al. (2002b) and the American Water Works Service Co. (AWWSC 2002) presented decision trees for the choices. The decision trees seemed almost identical, except that AWWSC added design factors such as numbers of connections and types of soil conditions.

In reviewing these decision trees, we concluded that the picture is more complex than can be depicted on a simple "either-or" type of display, and that a more detailed chart would be required to illustrate the decision paths. In Figure 7.9, we present a simple but general decision tree to illustrate the basic choices.

The decision tree illustrates that, based on the condition assessment, a pipe can have problems in structural integrity, hydraulic capacity, or water quality. Hydraulic capacity can be related to either flow or pressure. These problems can be interdependent; that is, a pipe can have one, two, or three of these problems. The problems divide into the six categories shown on the

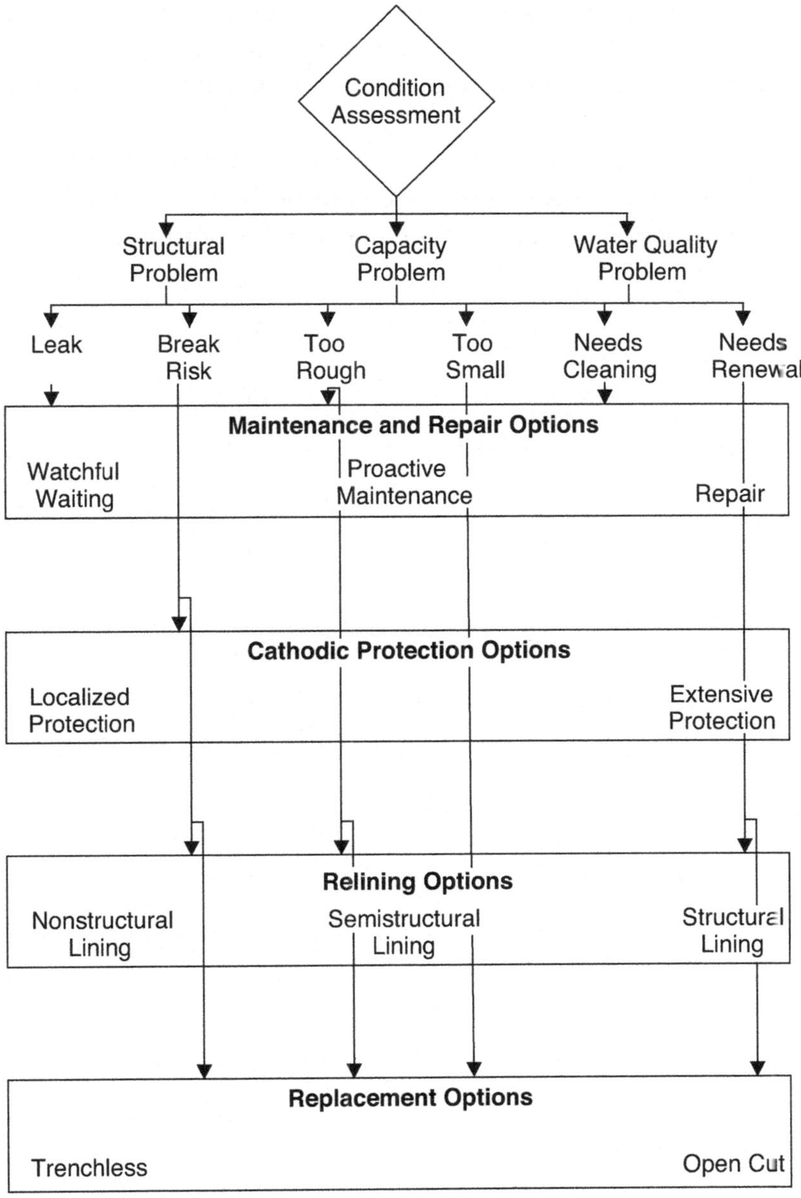

Figure 7.9 Decision tree for renewal projects

third line of the figure to remedy a water quality problem. This line illustrates the complexity facing the problem-solver at this point. The problem is neither linear nor either-or in nature; it is complex, requiring detailed analysis.

Each of the four types of solutions has ranges of investment. Maintenance and repair can range from "watchful waiting" (borrowing from medical terminology) to direct repair. Cathodic protection is somewhat unlike the other categories, in that it applies only to metal pipes and is a special type of solution. The term *relining* is used next instead of *rehabilitation*, because these solutions all involve some type of relining. If the project is cleaning and lining, as in CML or epoxy projects, cleaning must be done before lining.

Replacement involves a range of choices dealing with whether a pipe needs more capacity, whether special situations are involved (requiring trenchless construction), and other considerations. More work is required to prepare a design guide for utilities seeking to know which approach to use.

References we studied did not go into detail on how to evaluate choices, but a few choices were listed: soil conditions, number and type of connections, ease of excavation, aggressivity of water, congestion in construction area, and future plans and opportunities.

CHAPTER 8
UTILITY PROGRAM FOR DISTRIBUTION SYSTEM ASSET MANAGEMENT

INTRODUCTION

One of AwwaRF's requirements for this project report follows.

> The contractor will critically evaluate the information..., list the requirements for evaluating the various models..., and coordinate and moderate a workshop ... to review the current state-of-knowledge in the topic areas, identify remaining knowledge gaps, and develop a comprehensive plan to address the gaps.

AwwaRF's informal guidance was that utilities needed clarifying documents, as the research reports available to them tended to be overlapping and not always easy to follow.

This chapter responds directly to that requirement by briefly summarizing the tools available to utilities to use in planning and managing their distribution capital program.

BASIC TOOL KIT

The basic tool kit is the task map, as shown on Figure 1.1. It consists of

- Inventory as the core activity
- Condition assessment
- Maintenance program (preventive and corrective maintenance)
- System planning and needs assessment
- System financial accounting
- Asset management framework
- Capital improvement plan and program, including renewal technologies
- Capital budget
- Capital management system (strategy and framework)

Each of these tools has a research base, which is explained briefly in the following sections. The last section of the chapter describes how WSSC, one of the project's participating utilities, has undertaken the various tasks. WSSC's approach can serve as a model for other utilities facing similar situations.

Inventory as the Core Activity

The inventory is a set of maps, diagrams, and records that contain information about asset location and characteristics. At a minimum, a utility will have a set of these records and a plan to utilize and update the information.

An inventory can be as simple as, for example, a set of drawings indicating where pipes are located in a section of a city. Maintenance forces could use these drawings, with annotations, to locate and service pipes. On a more sophisticated level, the drawings could indicate other nearby facilities, such as sewer and electricity lines, and could also be used to coordinate services. Even more sophisticated would be coordinated data in a GIS and database format, available on a common basis to different sections of the organization that are responsible for shared data, processes, facilities, and staff.

Inventory is discussed generally in chapter 5, along with condition assessment. The important role of the utility database is stressed there. Several types of research-based technologies can be used in the inventory activity:

- Tools to locate assets
- GIS map systems
- Database management systems
- Handheld field devices

These tools are available commercially and, although their effectiveness may be advanced by research, for the most part they are technology-based devices that suppliers and consulting firms are constantly improving.

Condition Assessment

For condition assessment, the utility may have a proactive or reactive program. Because condition reporting is now required under GASB 34, it makes sense to organize a routine condition assessment and reporting activity that serves the entire utility. This activity provides information for planning maintenance, for the needs assessment and priorities, and for the capital program. See chapter 5 for a more detailed discussion of condition assessment.

Maintenance Program (Preventive and Corrective Maintenance)

The framework for organizing maintenance is the MMS, which brings activities together into a systems approach to investment, organization, scheduling, and monitoring. MMSs are described in different ways, but general functions are inventory, condition assessment, preventive maintenance, and corrective maintenance. Deb et al. (2000) reviewed MMSs.

In facilities management, the focus is increasingly on CMMS. According to Thomas (2001), writing in the *Facility Design and Management Handbook*,

> The effective implementation of a CMMS or work management program has become mandatory for facility departments. The implementation of a CMMS application represents one of the greatest challenges that a facility department will face, as the potential for radically changing the character of a facility department lies at the heart of a CMMS application.

Thomas considers the CMMS to be a sophisticated "to do list," and to have three parts: a preventive maintenance schedule, work order tracking, and project management.

System Planning and Needs Assessment

The capital improvement plan is the mechanism used to prioritize capital improvements, whether for new or renewed facilities. The *Capital Planning Strategy Manual* reviews knowledge about capital planning (AwwaRF, CH2M HILL, and AWWA 2002). The plan fits within an overall capital improvement program, which integrates capital needs and plans of all sections of the organization through the planning-programming-budgeting system (PPBS). Although a PPBS deals with budget and finance, it also has the wider role of integrating plans.

Early-stage planning, such as comprehensive or master plans, must be translated into specific plans for design and construction, then for O&M. Repair, rehabilitation, and replacement may reinitiate the planning cycle to complete the life-cycle approach to capital management. The capital improvement program thus can include facilities for growth, obsolescence, condition improvement, regulatory controls, or increases in reliability or service. It is easy to see the attractiveness of plans for new facilities, versus renewing older, invisible assets.

The *Capital Planning Strategy Manual* offers guidance for developing system plans, including priority setting. In chapter 6, we reviewed how different research products can be used to set priorities. The state of the art today begins with the basic step of collecting data to know something about needs, then proceeds to more sophisticated approaches. Utilities should choose from the following sequence:

- At the most basic level, have a list of the water mains most in need of attention, and choose from it each year to implement a capital plan for renewal. An important decision variable is the percentage of assets to renew each year.
- At the next level, improve the information content of the needs list by adding to it a scoring system where mains are rated according to need using criteria such as those discussed in chapter 6.
- At the most sophisticated level, use models such as KANEW and Nessie to maintain automated lists of investment needs and to use formal scoring systems to prioritize mains for renewal.

System Financial Accounting

A recurring theme in our project was that an effective renewal program requires using best practices in the financial arena. These include funding commitments, revenues, information use, and budgeting. To elaborate on these, the organization seeks sufficient funding and O&M costs are fully funded each year, alternative methods for financing are considered, agency requests are supported by solid information, and operating and capital budgets are linked (AWWA 1995, Grigg 2002).

One of the problems in renewal has been the neglected state of accounting for fixed assets as opposed to current assets (which have more dynamic financial turnovers). Actually, no comprehensive document has been published to bring together all the concepts of accounting for property, plant, and equipment (Peterson 1994). In fact, in government accounting, including water units, depreciation of fixed assets used to be optional. Now, with GASB 34, accounting for them is required.

The GASB, which specifies the Generally Accepted Accounting Principles for government accounting, has created a separate category for infrastructure fixed assets. Now, GASB requires government entities to report their capital assets in their annual balance sheet and income statement. GASB 34 was adopted in 1999 to identify the costs of acquiring, owning, operating, and maintaining infrastructure. It gives governments the choice of adopting traditional methods of calculating depreciation based on historical costs or setting up an asset management system.

Accounting practices, then, are one of the main constraints on doing a better job in capital management. Peterson (1994) asks:

> Do you have in place a process that monitors the current condition, evaluates the future need for replacement, and brings to your attention needs to modify that plan? . . . Government infrastructures of roads, sewers, sidewalks, and utilities are all suffering from the concept of "put it in place and forget about it."

Asset Management Framework

As we saw in Figure 1.1, asset management is the integration of the tools just described—inventory (as the core activity), condition assessment, maintenance program, system planning and needs assessment, and system financial accounting. Asset management can, of course, be described in different ways. It can also be called an infrastructure management system, which is an integrated framework for infrastructure through its life cycle from the cradle to the grave. No matter what term is used, the asset management system coordinates functions in an integrated, data-centered approach to managing an organization's physical systems. Asset management or infrastructure management systems are collections of fundamental management tools.

Capital Improvement Plan and Program

The capital improvement plan and program is, in the final analysis, a list of projects that an entity intends to implement under favorable circumstances. The uncertainty is about the intent and ultimate capability to follow though with the intent. Water system plans include facilities for growth, obsolescence, renewal, regulatory controls, and increases in reliability or service. In an ideal case, new systems can be completely financed by growth fees and developer contributions as "growth pays its own way." However, on a practical basis, the picture is more complex because the utility must share and participate in projects to ensure safety, coordination, and orderly growth.

AwwaRF's *Capital Planning Strategy Manual* (AwwaRF, CH2M HILL, and AWWA 2002) and AWWA's (1998) *Manual of Water Supply Practices* (M29) on water utility capital financing (including planning for capital improvements) offer guidance on developing the capital improvement plan and program. M29 explains strategic planning for capital requirements and financing mechanisms, mainly the bond market. Planning requires stakeholder interaction and leads to safety, reliability, and efficiency; annual cost minimization; optimal use of financial resources; and minimized annual impacts. This is a good statement of the financial part of planning for infrastructure, but the overall process involves many other considerations.

Engineering has an important part in capital improvement planning, as it selects the technologies for renewal (see chapter 7). In fact, the capital improvement plan starts with engineering's estimates of needs, then works up through prioritization, and finally to the capital program and budget.

Capital Budget

The capital budget flows from the capital plan and program. Capital expenses are incurred when building new systems or renewing old ones. New systems may be required for growth, improvement, or regulatory requirements. Old systems might require major repairs, rehabilitation, refurbishment, or replacement. Issues to address in capital budgeting include the time value of money, the cost of capital, the capital budgeting decision, and risk in capital budgeting.

Capital Management System (Strategy and Framework)

In the final analysis, a utility's capital management system consists of the array of fundamental tools summarized in this chapter. The system can be called by several names, but fundamentally, it means that the utility is giving proper attention to capital management, including distribution systems, through its management systems.

WSSC's APPROACH

The WSSC served as one of the participating utilities for the project and hosted a regional workshop early in the project. This provided an opportunity for WSSC to explain its comprehensive approach to water main analysis and renewal. Highlights are described briefly here to demonstrate how a utility program can organize and proceed.

WSSC has used modern techniques for condition assessment, and its Systems Infrastructure Group is now integrating a range of data into its decisions about pipe renewal activities (Woodcock 2002b). Chapter 5 describes these techniques briefly, including those applicable to CIP, DIP, and PCCP.

By the same token, WSSC has a comprehensive approach to planning and prioritization, as explained by Tucker and Corriveau (2003). As chapter 6 explains, this includes a prioritization algorithm that uses maintenance and break data to set priorities.

WSSC's approach to managing pipe data, including failure data, is shown by its maintenance records (dating to the 1970s), which are stored on a mainframe computer and made available through an information system to the UIMDSS. This is also described in chapter 6 and in Tucker and Corriveau (2003). WSSC also uses research products such as Nessie (see chapter 6) to display its needs.

Another way that WSSC illustrates good practices is in its long-term commitment to research about pipe infrastructure and renewal. The utility has published reports about its investigations since the 1980s and currently has information about metallurgy, failure mechanisms, and methods of condition assessment. In particular, WSSC has data about failure of CIP, DIP, and PCCP (Woodcock 2002b). For example, in some of their presentations utility managers display histories of break frequency versus water temperature, clearly demonstrating the linkage.

WSSC has also explained some of the research needed to improve our understanding of pipe behavior. For example, Woodcock (2002b) points out the need for research on detecting cracks in in-service pipes, evaluating old pipe without removing coatings, evaluating bed conditions, improving repairs such as backfill and injection stabilization, documenting the role of cracks in failures, creating a national database, and learning more about the future of DIP and polywrap.

CHAPTER 9
RESEARCH AND DEVELOPMENT NEEDS

INTRODUCTION AND PURPOSE

The purpose of this chapter is to synthesize reports about distribution system research needs, and to interpret them in the light of project findings. We also intend for the chapter to help outline a comprehensive program of research for consideration by the water supply industry for improving integrity of distribution systems.

This chapter begins by listing issue areas. Next, it presents a summary of research needs identified in the project. It then discusses the chronology of research initiatives for distribution systems, and concludes with suggestions for a research strategy.

CATEGORIES OF RESEARCH FOR DISTRIBUTION SYSTEMS

The issue categories are taken from chapter 2.

- Infrastructure management
- Planning, prioritization, and performance evaluation
- Pipe materials and structural research
- Condition assessment and failure mechanics
- Distribution system engineering, design and operation
- Maintenance management
- Construction and renewal
- Corrosion measurement and control
- Water quality maintenance and regulatory issues

GENERAL RESEARCH NEEDS IDENTIFIED DURING THE PROJECT

Most needs have been mentioned by multiple authors and in several workshops (Figure 9.1), so no attempt is made to attribute each one to individual sources. We have removed highly general statements from the list. The most strategic issues are listed in Table 9.1, and the full list is included as an appendix to this report.

CHRONOLOGY OF RESEARCH INITIATIVES

The water supply industry should consider its future research strategies in the context of how past research developed. The chronology of early research on distribution systems begins with a picture of science and technology during the 19th century. At that time, technologies were being developed in the areas of metallurgy, chemistry, biology, fluid mechanics, and civil engineering. AWWA, for example, was organized in 1881. The field of household plumbing evolved alongside the field of waterworks engineering. By 1900, knowledge about cast iron, pipe flow hydraulics, and structural mechanics was available. These developments form the early history of distribution system research.

Figure 9.1 Deliberations at project workshop

Early in the 20th century, new disciplines were being formed. During this period, mechanical, chemical, and sanitary engineering, along with the field of management, evolved. The invention of gasoline-powered construction equipment and electric power networks for pumping created new possibilities for water distribution systems. CIP standards were published in 1908, and drinking water standards began in 1914.

In this period, scientific journals and conference proceedings were not widespread, and research findings were reported only in textbooks. Early books about water supply and distribution offered guidance about pipe layout, sizing, installation, and other basic subjects. New valves, fittings, and other components of distribution systems were being developed. By the 1930s, plastic pipe had emerged in Germany, and by the 1950s, concrete pipe was in use. The field of pipeline engineering began to evolve in this period. By the 1940s, practice-oriented papers were appearing in *Journal AWWA* to report on distribution system integrity, mainly about CIP.

Research supported by federal agencies in the 1960s and 1970s focused on pipe-related subjects. USEPA was studying distribution systems during the 1980s, and the U.S. Army Corps of Engineers was conducting a project on distribution systems at the Construction Engineering Research Laboratory, which also focused on condition assessment systems for infrastructure. Meanwhile, the discipline of pipeline engineering had emerged, and its adherents were discussing research needs. The Pipeline Division of ASCE was authorized in 1956, and in 1987, with the support of the National Science Foundation (NSF), the division published a report on the research needs of that time (Selig and Nash 1988). Thirty-five participants from universities, consulting engineering and research firms, pipe and equipment manufacturers, constructors, pipe users, and

Table 9.1
Strategic distribution system issues

Category	Research Issue
Infrastructure management	• Institutional issues to include promoting innovation, policy for public investment, overcoming barriers to new technologies, and a regulatory framework for buried infrastructure management • Funding mechanisms and procedures to improve rate of renewal of systems • Technology transfer and workforce development • Organizational systems to handle the large amount of data and analysis required for distribution systems capital management
Planning, prioritization, and performance evaluation	• Utility databases to study costs of maintenance, repair, rehabilitation and new construction • Guidelines for determining which pipes should be tested or replaced • Improved knowledge for predicting service life • Method to build case for pipeline renewal
Pipe materials and structural research	• Structural behavior of all pipes, plastic pipe properties, and verification studies of failure mechanisms for all pipe types (continue long-term research program)
Condition assessment and failure mechanics	• Development of a national database of failures, causes, and renewal methods, including standard reporting methods • Continued work on failure mechanisms with different types of pipe; causes for pipe, joint, lining, and coating deterioration • Continued work on methods for inspecting and assessing all types of pipes • Asset location (emphasizing finding small things and nonmetallic assets) • Smart pigs to locate adjacent buried infrastructure • Continued development of instrument and instrument systems to include smaller hydrophones, permanent transducers, and smart implanted chips • Improved pig for all purposes in water distribution systems
Distribution system engineering, design and operation	• Codification of design information • Improved specifications for difficult installation conditions • Distribution system performance standards • Evaluation of long-term performance of CML • Design manual with case studies of special conditions and decisions, with why these decisions were made
Maintenance management	• Best management practices for distribution systems
Construction and renewal	• Installation of service connections • Lining techniques for return to service in same day • Installation to bond continuity cables across pipe joints
Corrosion measurement and control	• Outside scope of project, but continued strategic program of corrosion and protection research is required
Water quality maintenance and regulatory issues	• Outside scope of project, but continued strategic program of research on water quality and distribution systems is required

government agencies contributed to the Selig and Nash report. It contains a long list of pipeline engineering subjects that required research at the time.

After AwwaRF was organized in the 1980s, research on distribution systems increased. *Journal AWWA* also began to include periodic theme issues focused on distribution systems. The report by O'Day et al. (1986) contained information about pipe integrity research needs. AwwaRF's (1989) proceedings on research included discussions of the following issues: conclusions of the WSSC about the need for statistics on breaks as well as findings on pipe material and soil condition, a natural gas industry infrastructure overview, case studies of projects using automated maps, review of trenchless technologies, location of pipe and leaks using GPR, NDT, one-call systems, corrosion of services, cathodic protection, and a national perspective on infrastructure costs. This report set the stage for an AwwaRF project to identify research needs.

AwwaRF's report by Kirmeyer, Richards, and Smith (1994) compiles knowledge through the early 1990s. The report by Smith et al. (2000) and Ellison's (2001) AwwaRF report bring some of the topics up to date and add new ones. Meanwhile, other groups, such as USEPA, continue to conduct research. All in all, a great amount has been learned about distribution systems in the last few decades.

TOWARD NEW RESEARCH STRATEGIES

Given the scale of the challenge of managing distribution systems, a new strategy is required. The strategy should consider the future of the utility business, the realities of the distribution system's underground environment, and the knowledge base that has been accumulated.

The Future of the Utility Business

A number of writers have peered into the future of water utilities. Most seem to predict that utilities will be more businesslike; more privatized and/or consolidated; more regulated, with greater scrutiny on health and water quality; more conserving of water; and more focused on systems integration and management. In addition, some predict that utilities will have higher rates and a workforce that is more skilled in technology, with greater use of information systems. Some forecast a new kind of utility that offers bundled services in utility and business sectors, sort of a utility Wal-Mart.

Shorney (2000) lists forces that are transforming the water business: public health, public affairs, globalization, privatization and outsourcing, competition, total quality management, accreditation, strategic planning, deregulation, financial practices, utilities and capital, rates and charges, resource planning and management, demand management and water efficiency, regulatory affairs, customer service, public involvement, workforce, compensation and benefits, operator training, information technology, project delivery, and capacity. It is interesting that the problem of distribution system capital did not make the list.

The Realities of the Underground Environment

To face the reality of the underground infrastructure environment, we start with the knowledge that the current pipe inventory is mostly CIP or DIP. Given the current replacement rate of about 1 in 200, in the next 50 years we can expect gradual replacement of gray cast iron with ductile iron and other materials such as PVC, HDPE, concrete, and steel. By 2050, there will be more pipe, with changed characteristics, but the inventory will still be mainly iron pipe (although the percentage of cast iron will be less).

The revolution in distribution system integrity could come in condition assessment. By projecting the technological advances of the last 25 years into the future, we see that our ability to see underground; to sense pipe faults; to locate pipes, leaks, and breaks; and to monitor water quality will increase greatly. The frontier will be economic. It is one thing to spend $100,000 on the health care of an individual, and it is quite another thing to spend hundreds of thousands of dollars on sensing the condition of underground pipes. The value of the inspections has not been established yet, but it will be. Water is becoming more precious, health aspects are coming under more scrutiny, security is becoming a bigger issue, and all other trends are pointing that way. We can expect many innovations.

The Knowledge Base

Technologies and techniques for planning will continue to evolve. The most sophisticated utilities will develop their mapping and information systems to a greater degree and will demonstrate a high level of skill in planning. The most dramatic change will be the extent to which sophisticated planning techniques diffuse to other utilities. By 2050, most utilities will have a water mains database, a formal condition assessment program, and employees with good integrated senses of pipes needing renewal. They will use some management science techniques, such as optimization and risk analysis, but these will still be used as guides, not to actually make the decisions.

Future Trends

The main trend in the future will be integrating activities, or "systems integration and management," as described in AWWA's publication, *The Changing Water Utility* (Westerhoff et al. 1998).

Of the future trends in the utility business, several stand out as influencing how distribution systems should be managed. These include being more businesslike, setting rates to cover full costs, employing smaller and even more skilled workforces, using information systems more often, and being more regulated in terms of health and water quality.

Regardless of future pressures on water utilities, only two alternative futures for water distributions systems seem likely. In the most likely scenario, the distribution system will look much like it does today, with many miles of main that are 8 in. and larger, comprising different materials, fittings, and patches. Treated water would be potable, and utilities would be regulated under the SDWA of 1974, with increasingly strong controls brought on by amendments and rules. Although concerns about deteriorating water quality in distribution systems would remain, few health problems would be reported. For that reason, utilities would be able to continue operations much as they have for the past 50 years. Main breaks and leaks would be reduced greatly with the vast improvement of leak detection and condition assessment methods. It is true, of course, that people may never trust tap water, so the bottled water industry and point-of-use industries would prosper.

In a second, more hypothetical scenario, utilities would be under more and more pressure from about 2010 to 2025, when the government would pass new rules to establish the right to drinking water of the purest quality for each American. Previous laws and rules would have been adequate until 2010, but utilities would come to realize that they could not guarantee the safety and security of potable water from each and every point of their vast networks with practical means. For that reason, the government would establish a new category of water supply, "treated pure water," to accompany the existing categories of raw water and treated water. Treated water would remain potable at the treatment plant, and utilities would continue to monitor and control water quality in distribution systems, but the end user would also be supplied with additional safety from zonal treatment systems to polish and monitor the water before it enters metal or plastic final piping systems (which would be attached to home plumbing systems). Utilities would continue to manage distribution systems much as they did in the past, but they would put many more monitors, safeguards, and protective systems in place.

Under either of these scenarios, the management of distribution systems would be similar, although in the second, more stringent strategy, additional requirements would result.

A General Strategy

Kirmeyer, Richards, and Smith (1994) proposed a vision statement for the strategic plan:

> The strategic goal for AwwaRF in the area of Distribution Systems is to conduct research that will enable utilities to plan, design, construct, operate, and maintain transmission and distribution systems in a manner that (1) minimizes water quality deterioration through to the customers' taps and (2) extends the life of and makes efficient use of pumping plants, storage facilities, conveyance systems, and water system appurtenances.

Any vision statement can be improved, but this one contained the basic elements required for that time. Perhaps it should be modified now to place more emphasis on business operations and system renewal.

Kirmeyer, Richards, and Smith's (1994) recommendations to organize the categories of distribution systems research are summarized here. Most of the topics addressed in their report are lumped into one category, water distribution system infrastructure. Certainly the other categories will continue to be essential, so one option for organizing the research on infrastructure is to subdivide it, possibly using the categories presented in the Kirmeyer report: management; planning, prioritization, and performance evaluation; pipe materials and structural research; condition assessment and failure mechanics; distribution system engineering, design and operation; data management and distribution system modeling; maintenance management; and construction and renewal. These would go along with water quality maintenance, regulatory issues, corrosion, permeation, and materials deterioration to form a longer list of subtopics.

Since the publication of Kirmeyer, Richards, and Smith's (1994) report (summarized in Table 9.2), AwwaRF has funded a number of research studies that relate directly to its recommendations. AwwaRF's (2001) report summarized projects completed as of that date, and a list of more recent projects is included in the next section.

Current AwwaRF Research on Distribution Systems

Table 9.3 summarizes current AwwaRF research, including RFPs. Projects are listed by categories as used in this report. Note the categorization system.

Table 9.2
Recommendations for program structure

AwwaRF category	Recommendations
Prevention of water quality deterioration	Change name to "Water Quality Maintenance and Regulatory Issues." Change focus to respond to regulatory changes. Study causes of water quality deterioration in distribution systems and practical solutions. Emphasize bacterial regrowth, DBPs, residuals, lead service lines, and taste and odor control.
Corrosion measurement and control	Change name to "Corrosion, Permeation, and Materials Deterioration." (The authors thought that the LCR would shift emphasis, and that a change from lab to field research would be needed.) Study causes and solutions for corrosion, materials deterioration, leaching, and permeation. Emphasize corrosion and leaching, movement of contaminants through nonmetallic pipe and joints, and bacterial regrowth interaction with corrosion.
Infrastructure	Change name to "Water Distribution System Infrastructure." (They saw extensive needs for more work and thought that projects should focus on better testing, evaluation, and methods for new pipe and renewal work.) Stress applications research for test methods and renewal field methods. Emphasize NDT, rehabilitation of unlined pipes, test methods, and external corrosion. In this category, the report recommended larger projects with more direction from AwwaRF staff.
Distribution system modeling	Change name to "Distribution System Design and Operation." Stress tools to locate and assess condition of system components. Emphasize performance standards, control systems, models, and mapping.

Source: Kirmeyer, Richards, and Smith 1994

The current count of projects is given in Table 9.4. Obviously, condition assessment and water quality are deemed to have the most research needs. Only one project on security is listed, but quite a few other projects on security have been initiated. They are not always included in AwwaRF's listings for security reasons.

RESEARCH POLICIES

Naturally, the research policies of AwwaRF and the water supply industry must be comprehensive and consider more topics than distribution systems, which in and of themselves present formidable challenges. A few contextual realities must also be considered, such as

- The water supply industry is a broad sector that deals with natural water systems, public health, infrastructure, public–private management issues, and many others.

Table 9.3
Current AwwaRF research on distribution systems*

C: Optimizing Corrosion Control in Water Distribution Systems (2648)

C: Identify Causes and Solutions for External Corrosion on Buried Water Mains (2608)

CA, P, R: Synthesis Documents on Infrastructure Integrity (2772)

CA: Life-Cycle Analysis of Epoxy Lining (2941)

CA: An Investigation of Remote Sensing Methods for Leak Detection in Rural Water Transmission Pipelines (2826)

CA: Circumferential Failures in Gray Cast Iron Distribution Pipes (2727)

CA: Development of Non-Interruptive Condition Assessment Inspection Devices for Water Transmission Mains (2871)

CA: Evaluate Petroleum-Based Hydrocarbons on PVC and PE Pipe (2946)

CA: Evaluating Water Loss and Planning for Loss Reduction Strategies (2811)

CA: Impact of Petroleum-Based Hydrocarbons on PE/PVC Pipes and Pipe Gaskets (2946)

CA: Leakage Management Technologies (2928)

CA: Life-Cycle Analysis of Epoxy Lining (2941)

CA: Long-Term Performance Prediction for Polyvinyl Chloride Pipe (2879)

CA: Multi-Utility Buried Pipes and Appurtenances Location Workshop (2882)

CA: Performance and Life Expectancy of Elastomeric Gaskets in Contact With Potable Water (2932)

CA: Techniques for Monitoring Structural Behavior of Piping Systems (2612)

CA: Testing and Condition Assessment of Joints in Water Distribution Pipelines (2689)

CA: Water Supply System Performance in the Nisqually Earthquake (2846)

E: Impacts of Fire Flow on Distribution System Water Quality, Design, and Operation (2554)

E: Criteria for Valve Location and System Reliability to Optimize Continuity of Water Supply (2869)

E: Development of Distribution System Optimization Plans (2875)

E: Proceedings of a Workshop on Seismic Measures for Water Supply (2786)

P: Risk Management of Large Water Transmission Mains (2883)

P: Customer Acceptance of Infrastructure Reliability (2870)

R: Cost–Benefit Analysis of Flushing (2605)

R: Evaluate How Reliability-Centered Maintenance Can Be Implemented in Water Utilities (2953)

R: Evaluate Ways to Minimize Downtime During Pipe Lining Operations (2956)

R: Non-Disruptive Tools for Remaking Connections After Pipe Rehabilitation (2872)

S: Vulnerable Points in Water Distribution Systems (2931)

WQ: Physical and Numerical Modeling of Mixing in Water Storage Tanks (2898)

(continued)

Table 9.3 (Continued)

WQ: Study Water Quality Changes in Distribution Systems as a Result of New Development and Low Use (2954)
WQ: A Rapid Ammonia-Oxidizing Bacteria Measurement Method as an Early-Warning Indicator of Nitrification Episodes (2899)
WQ: Distribution Water Quality Issues Related to New Development or Low Usage (2954)
WQ: Evaluate Point-Of-Use Systems to Assess Extent of Microbial Contamination in Distribution Systems (2986)
WQ: Evaluating Retention Time to Manage Distribution System Water Quality (2769)
WQ: Impacts of Changes in Distribution System Water Quality on Disinfection Efficacy (2771)
WQ: Improving Methods for Classification of Exposure to DBPs (2831)
WQ: Long-Term Effects of Disinfection Changes on Water Quality (2940)
WQ: Predictive Models for Water Quality in Distribution Systems (2865)
WQ: Qualitative Procedures for Identifying Particulate Matter in Distribution Waters and at the Customer's Tap (2866)
WQ: The Formation and Decay of Disinfection By-Products in the Distribution System (2770)
WQ: Workshop to Develop Methods to Evaluate Health Risks Associated With Infrastructure Failure (2773)

*Categories are as follows:
C = corrosion
CA = condition assessment and failure mechanics
E = distribution system engineering, design and operation
P = planning, prioritization, and performance evaluation
R = construction and renewal maintenance management
S = security (added topic)
WQ = water quality maintenance and regulatory issues

Table 9.4
Project count

Project	Total Number
Management and synthesis (S)	1
Corrosion (C)	2
Condition assessment (CA)	16
Engineering (E)	4
Planning (P)	3
Maintenance and renewal (R)	5
Water quality (WQ)	13

- AwwaRF is the main U.S. coordinating organization for research conducted in the water supply industry.
- Many collaborating organizations perform related research, including utilities, water supply associations, and other countries.
- Utilities are heavily regulated and need a broad knowledge base for regulatory compliance and business operations.
- The water supply industry's commercial sector is able to produce many innovations to respond to incentives.
- Many scientific and business organizations are available to conduct research for the water supply industry.

It was in this same context that the Kirmeyer, Richards, and Smith (1994) report recommended policy changes for research into distribution systems. They suggested placing more emphasis on public–private cooperative efforts; identifying universities to specialize in the rehabilitation of water systems; increasing international cooperation and technology transfer; conducting larger and longer duration projects stressing field work, not surveys and paper studies; and developing new project selection and management approaches for prototype research projects. These recommendations focus on the cooperative nature of the industry and point to a need to take projects into the demonstration stage.

Even though AwwaRF should focus on strategic issues, it still serves the practical needs of utilities with questions such as those that led to this project: Should utilities be reactive or proactive in renewing distribution systems? How can utilities get funding for system renewal among so many competing priorities? How should utilities prioritize renewal projects? Given its complexity, how should utilities manage data on system failures and renewal? Should utilities simply continue to repair old systems, or should they renew them? To what extent should utilities use new trenchless technologies?

As the nation developed its federally sponsored research programs, such as those in the NSF, U.S. Department of Defense, U.S. Department of Energy, and others, it created a system for research management that still offers a model for publicly supported research programs. The model has many forms, but it generally consists of research need identification, basic research, applied research, demonstration projects, technology transfer, and training. AwwaRF's program includes these phases, but it cannot handle all of them fully, both because of their scope and because of its partners' activities.

It was not in the scope of this project to study research strategy documents, but AwwaRF's strategic research plan (AwwaRF 2004) emphasizes practical, applied projects, and basic research is conducted to lead to practical, applied tools and technologies. The four strategic research goals—an efficient and customer-responsive organization, environmental leadership, high-quality water, and infrastructure reliability—shape the program. Distribution system projects relate to all four goals, especially to infrastructure reliability and to high-quality water.

When we polled our project participants at the workshop to get their recommendations about changed research topics and priorities, they identified some that dealt with strategy: coordinate research more to remove duplication and overlap; expand the reach of reports beyond utility top managers; make results more user friendly; release reports faster, but still with good review; improve Web-based capabilities to find applications of interest; and consider an industry research institute that does R&D in house, like GTI in the gas industry. Although AwwaRF has probably heard these suggestions before, and they have not been tested on broader stakeholder groups, they still contain valid suggestions for areas of emphasis and study.

By selecting a few areas of emphasis from this project, strategic needs are seen clearly. For example:

- Infrastructure management—to make progress on institutional issues to promote innovation, to set policy for public investment, to overcome barriers to new technologies, and to develop an improved regulatory framework for buried infrastructure management
- Planning, prioritization, and performance evaluation—to stimulate research and testing on utility databases for pipe management and to develop a national database of failures, causes, and renewal methods, including standard reporting methods
- Condition assessment—to conduct continued research on failure mechanisms with different types of pipe; to determine causes for pipe, joint, lining, and coating deterioration; and to continue integrating results, with guidance for utilities in this complex arena
- Distribution system engineering, design and operation—to codify design information
- Renewal—to install service connections more effectively and to continue evaluating renewal options

If a single finding stands out, it is that a process is needed to periodically bring together research findings in focused areas and publish the condensed findings in ways that are helpful to utilities. At the same time, AwwaRF should continue to identify needs and carry out applied research until the findings are delivered to the users. For distribution systems, the area of condition assessment stands out as most in need of further classification, which would lead to publishing condensed reports in applied areas.

CONCLUSIONS

The many topics addressed in this report illustrate the breadth of the water pipeline research field. Perhaps this quote from a 15-year-old report best captures the spirit of the issue (Selig and Nash 1988):

> Buried pipe infrastructure by its very nature is not a highly visible field of engineering, and not one of major public awareness or glamour. The participants are often relatively small companies or government agencies. For these reasons the extent of technology growth as represented by research activity has been relatively small. An important conclusion from the workshop is that a greatly increased research effort is both needed and justified by economic considerations.

APPENDIX
RESEARCH NEEDS LIST

INFRASTRUCTURE MANAGEMENT

Research needs were identified in policy, knowledge management, and management methods.

Policy

- Promote innovation, establishing policy for public investment to speed development and incentives for private sector to adapt technology.
- Identify institutional barriers to new technologies.
- Increase confidence of water utility managers to use new renewal technologies.
- Set goals for ideal distribution system of the future.
- Develop guidelines and standards.
- Develop regulatory framework for buried infrastructure management.
- Promote university attention to rehabilitation and renewal, more public–private efforts, and more international cooperation.
- Create research foundation similar to oil and gas industry to implement technology.
- Find appropriate ways to fund pipe renewal programs.

Knowledge Management

- Evaluate technologies and learn from failures.
- Disseminate information.
- Train workforce.
- Review European practices.
- Identify critical factors for successful projects.

Management Methods

- Develop methods to manage pipeline renewal programs.
- Identify best methods to handle substantial data and sophisticated analysis.

PLANNING, PRIORITIZATION, AND PERFORMANCE EVALUATION

- Create industry standard for determining which pipes should be tested or replaced.
- Develop criteria for prioritization.
- Identify critical versus noncritical links.
- Predict service life.
- Study properties of pipes that have been in long periods of service.
- Study methods to extend life.

- Develop methodology for decisions.
- Determine probability and consequences of failure.
- Develop model for deciding among alternatives, right level of renewal—optimization.
- Develop economic assessment methodologies.
- Develop multiattribute pipe condition ranking.
- Produce database on costs of maintenance, repair, rehabilitation, and new construction.
- Improve data reliability for decisions.
- Develop a small-utilities prioritization model.
- Prepare guidelines for geographic information system (GIS) development/implementation/use and lessons learned.
- Develop a decision support model (life-cycle and financial model).
- Develop a method to build a case for pipeline renewal.

PIPE MATERIALS AND STRUCTURAL RESEARCH

- Structural behavior: Determine longitudinal effects, buckling, time-dependent plastic pipe properties, and reinforced concrete pipe tension strength.
- Plastic pipe research: Determine the performance of polyvinyl chloride, size effects, fracture toughness, stress concentrations, minimum stiffness, and flexible pipe re-rounding. Perform a condition assessment of polymeric pipe.
- Metallic pipes: Conduct further verification studies.
- Asbestos–cement (A–C) pipe: Develop guidelines for management, including policies for leaving or removing it.
- Develop methods to inspect A–C pipe.
- Conduct further testing of prestressed concrete cylinder pipe (PCCP) with destructive evaluation.
- Conduct a study of whether polywrapped ductile iron is as corrosion-resistant as pipe with bonded coatings.

CONDITION ASSESSMENT AND FAILURE MECHANICS

In general, the goals are to develop more accurate, user-friendly test methods to determine condition of pipe; to expand understanding of causes for deterioration, leaks, and breaks; and to prevent problems and predict length of life under various conditions.

Failure Mechanics

- Encourage better utility records and standardized main break databases for utilities.
- Provide a national database of failures, causes, and renewal methods.
- Develop standards for pipe performance and failure histories.
- Create a standard form to report failures, such as the Gas Technology Institute's.
- Determine a workable way to predict when individual pipes need renewal.
- Determine failure mechanisms with different types of pipe and causes for pipe, joint, lining, and coating deterioration.
- Examine corrosion failure modes, internal corrosion, and external corrosion.

Methods

- Inspect and assess pipe of the future: polymer-type pipes.
- Perform condition assessment for critical mains.
- Perform nondestructive evaluation (NDE) for concrete pipe.
- Develop standard procedures for controlled destructive evaluation for use in large-diameter steel pipe.
- Develop guidance for supervisory control and data acquisition systems relating to condition assessment.
- Develop methods to monitor corrosion and determine effectiveness of protection.
- Conduct testing of PCCP with destructive evaluation.
- Perform structural monitoring.

Leak Detection

- Develop industry standards for "water loss" similar to those in the United Kingdom.
- Improve leak detection equipment to reduce noise interference.
- Develop a method for zone metering for leak detection.
- Develop a method for leak detection in large-diameter pipes using neutrally buoyant hydrophones.

Location

- Capture and record data; develop data sharing techniques and standards for asset location data collection and exchange.
- Create ways to find small items and nonmetallic assets.
- Evaluate equipment performance.
- Use smart technology, ground-penetrating-radar performance, and smart pigs to locate adjacent buried infrastructure; horizontal sensing techniques for determining depth; and "see ahead" technologies and methods for detecting nonferrous buried assets.
- Use asset tagging.
- Develop a multisystem location system.
- Categorize the utility environment.

Instruments

- Use smaller hydrophones to fit through blow-off valves.
- Apply remote field eddy current (RFEC) to reinforced concrete pipe.
- Use permanent transducers on transmission pipelines with remote monitoring.
- Use magnetic flux leakage devices.
- Use an RFEC from oil and gas.

Pigs

- Use pigs to go around sharp bends.
- Develop piggable (fully open) valves.

- Use launching and catching devices for NDE equipment into water pipes.
- Use smart pigs.

DISTRIBUTION SYSTEM ENGINEERING, DESIGN AND OPERATION

- Determine design information: performance limits, safety factors, loading and soil pipe interaction, bedding conditions (influence of bedding on pipe load capacity), three-dimensional effects, soil parameters, loads on pipes, lateral and longitudinal restraints, interface shear resistance, time-dependent effects, buckling, and post-yield performance.
- Develop probabilistic models and determine risk assessment.
- Develop expert systems for design.
- Develop installation methods and conditions for difficult installation conditions.
- Perform damage prevention.
- Develop procedures for thawing frozen ground.
- Conduct test procedures for new pipe.
- Develop remedies for pipeline materials with poor performance.
- Develop distribution system performance standards.
- Develop guidance for mapping and GIS systems.
- Develop guidelines for models in system design and operation.
- Determine coordination for replacing pipe when streets are renewed.
- Perform experimental verification.
- Determine seismic performance.
- Determine design issues such as contracting and liability.
- Determine long-term performance of field-applied cement–mortar lining.

MAINTENANCE MANAGEMENT

- Maintain emergency shutdown systems.
- Determine system flushing needs, monitoring, and impact on water quality.
- Develop leak-response benchmarks.
- Determine maintenance effectiveness.
- Develop best management practices for distribution systems.
- Determine ways to increase system life.
- Develop effective in situ tapping methods.
- Develop methods for distribution system evaluation.

CONSTRUCTION AND RENEWAL

- Perform installation of service connections.
- Develop lining techniques for same-day return to service.
- Install flexible pipe.
- Perform bedding construction (consider haunch compaction techniques, choices of materials, effects of low compaction, and prevention of fine soil migration).
- Conduct compaction testing.
- Perform trench bracing.

- Develop equipment automation.
- Determine loads in poor soil conditions.
- Develop installation technique for bonding electrical continuity cables across pipe joints using vacuum excavation.
- Conduct spot repair and replacement.
- Determine contracting methods, construction techniques and methods for specific tasks, such as excavation and trenching, bedding and pipe laying, and backfilling.
- Construct new machinery and use trenchless technologies.

CORROSION MEASUREMENT AND CONTROL

Corrosion research was generally outside of the scope of this project, but we identified general aspects that are centrally related to pipe integrity:

- Concrete pipe corrosion protection systems
- Internal corrosion and loss of wall thickness
- Cathodic protection evaluation
- Standards of performance for corrosion control
- Deterioration and leaching of pipe linings and coatings
- Permeation of pipe and joint materials by solvents
- Causes of and preventive measures for external corrosion
- Effects of grounding electrical services to pipe

WATER QUALITY MAINTENANCE AND REGULATORY ISSUES

Water quality research was outside the scope of the project. In general, research needs cited in the reports we reviewed covered technical issues about contaminants of concern, both from regulatory and security standpoints. New thinking about water quality monitoring and modeling is needed to merge regulatory and security concerns. Needs include type of sensors, siting parameters, data processing principles, and ways to control false positives and negatives. Research might identify a combination of condition and contaminant monitoring that utilities could implement.

REFERENCES

Allen, M.J. 1985. Developing Realistic Replacement-Rehabilitation Criteria for Distribution System Components. In *Proceedings of the Distribution System Symposium*, September 8–11, 1985. 191–194. Denver, Colo.: AWWA.

Anderson, R.B. 1995. Anatomy of a Pipeline Break. *Jour. New England Water Works Assn.*, 109(3):188–191.

Andreou, S.A., D.H. Marks, and R.M. Clark. 1987. A New Methodology for Modeling Break Failure Patterns in Deteriorating Water Distribution Systems: Theory. In *Advances in Water Resources*, 10(1):2–10. Amsterdam: Elsevier B.V.

Arnold, G.E., E.J. Clark, G.J. Remus, and H.W. Niemeyer. 1960. Experience With Main Breaks in Four Large Cities. *Jour. AWWA*, 52(8):1041–1058.

ASCE (American Society of Civil Engineers). 2001. Report Card for America's Infrastructure [Online]. Available: <www.asce.org/reportcard>. [cited February 13, 2002]

Australian Academy of Technological Sciences and Engineering. 2003. Technology in Australia 1788–1988. Cement-Fibre Pipes [Online]. Available: <www.austehc.unimelb.edu.au/tia/846.html>. [cited June 30, 2003]

AWWA (American Water Works Association). 1995. *Water Utility Accounting*. 3rd ed. Denver, Colo.: AWWA.

AWWA (American Water Works Association). 1997. *Water:\Stats—The Water Utility Database*. 1997 version [CD-ROM]. Denver, Colo.: AWWA.

AWWA (American Water Works Association). 1998. Water Utility Capital Financing. In *Manual of Water Supply Practices*. M29. Denver, Colo.: AWWA.

AWWA (American Water Works Association). 1999. *Installation of Ductile-Iron Water Mains and Their Appurtenances*. AWWA C600-99. Denver, Colo.: AWWA.

AWWA (American Water Works Association). 2001. *Rehabilitation of Water Mains*. M28. Denver, Colo.: AWWA.

AWWA (American Water Works Association). 2002. *Water:\Stats—Distribution System Survey* [CD-ROM]. Denver, Colo.: AWWA.

AWWA (American Water Works Association). 2003. *Water Transmission and Distribution*. 3rd ed. Denver, Colo.: AWWA.

AWWA (American Water Works Association). 2004. Officers and Committee Directory: AWWA Volunteer Units. Water Loss Control [Online]. Available: <http://www.awwa.org/about/oandc/committee/index.cfm?vid=00500385&cid=00500364>.

AWWA Water Loss Control Committee. 2003. Applying Worldwide BMPs in Water Loss Control. *Jour. AWWA*, 95(8):65–79.

AwwaRF (Awwa Research Foundation). 1981. Infrared Scans at Four Michigan Cities. *AwwaRF Water Qual. Res. News*, (November)26:2–4.

AwwaRF (Awwa Research Foundation). 1989. *Proceedings for the First International Conference on Underground Infrastructure Research*. Denver, Colo.: AwwaRF.

AwwaRF (Awwa Research Foundation). 2001. Synthesis Documents on Infrastructure Integrity. RFP 2772. Denver, Colo.: AwwaRF.

AwwaRF (Awwa Research Foundation). 2004. Strategic Research Plan. [Online]. Available: <http://www.awwarf.org/theFoundation/ourPrograms/strategicResearchPlan.aspx>. [cited January 15, 2004]

AwwaRF (Awwa Research Foundation), CH2M HILL, and AWWA (American Water Works Association). 2002. *Capital Planning Strategy Manual*. Denver, Colo.: AwwaRF.

AwwaRF (Awwa Research Foundation) and DVGW-Technologiezentrum Wasser. 1996. *Internal Corrosion of Water Distribution Systems*. 2nd ed. Denver, Colo.: AwwaRF.

AwwaRF (Awwa Research Foundation) and Japan Water Works Association. 2001. *Proceedings of the 2nd Japan and U.S. Workshop on Seismic Measures for Water Supply*. Denver, Colo.: AwwaRF.

AwwaRF (Awwa Research Foundation) and U.K. Water Research Ltd. 2002. Multi-Utility Buried Pipes and Appurtenances Location Workshop, London, May 29–31, 2002.

AWWSC (American Water Works Service Co., Inc.). 2002. Deteriorating Buried Infrastructure: Management Challenges and Strategies. White Paper [Online]. Available: <http://www.epa.gov/safewater/tcr/pdf/infrastructure.pdf>. [cited August 11, 2003]

Baracos, A., W.D. Hurst, and R.F. Legget. 1955. Effects of Physical Environment on Cast-Iron Pipe. *Jour. AWWA*, 47(12):1195–1206.

Becker, J. 2003. Water Distribution System Assessment and Hydraulic Model. Sewerage and Water Board of New Orleans. Presented at Distribution System Infrastructure Integrity Workshop, April 3–4, at Colorado State University, Fort Collins.

Bianchetti, A.W. 2001. *Peabody's Control of Pipeline Corrosion*. Houston, Texas: National Association of Corrosion Engineers.

Booth, R., and J. Rogers. 2001. Using GIS Technology to Manage Infrastructure Capital Assets. *Jour. AWWA*, 93(11):62–68.

Boyle Engineering Company. 2000. *Investigation of Pipe-Cleaning Techniques*. Denver, Colo.: AwwaRF.

Bradley, L.C., Jr., and F. Crumb. 1987. Confluence—Ridding Pipe of Persistent Joint Leaks. *Jour. AWWA*, 79(10):96–97.

Braun, E. 1998. *Technology in Context: Technology Assessment for Managers*. New York: Routledge Publishers.

Brown, W.G., ed. 1998. *Nondestructive Evaluation of Utilities and Pipelines II*. Bellingham, Wash.: International Society for Optical Engineering.

Buie, L.M. 2000. Accounting for Lost Water. *Jour. AWWA*, 92(7):67–71.

Cabrera, E., A. Miguel, and P. Francisco. 1995. Network Maintenance Through Analysis of the Cost of Water. *Jour. AWWA*, 87(7):86–98.

Cates, W.H. 1971. *History of Steel Water Pipe: Its Fabrication and Design Development*. Publisher unknown.

Cesario, L. 1995. *Modeling, Analysis, and Design of Water Distribution Systems*. Denver, Colo.: AWWA.

CIRIA (Construction Industry Research & Information Association). 1996. *Water Mains: Guidance on Assessment and Inspection Techniques*. London: CIRIA.

Clark, R.M., and W.M. Grayman. 1998. *Modeling Water Quality in Drinking Water Distribution Systems*. Denver, Colo.: AwwaRF.

Clark, R.M., M. Sivaganesan, A. Selvakumar, and V. Sethi. 2002. Cost Models for Water Supply Distribution Systems. *J. Water Res. Pl.-ASCE,* 128(5):312–321.

Clift, J.S. 2001. *A Legacy of Pipe*. Dayton, Ohio: Price Brothers.

Clinton, P. 2003. EBMUD Pipeline Replacement Program. Presented at Distribution System Infrastructure Integrity Workshop, April 3–4, at Colorado State University, Fort Collins.

Comptroller General of the United States. 1980. *Additional Federal Aid for Urban Water Distribution Systems Should Wait Until Needs Are Clearly Established.* (November 24) CED-81-17. Washington, D.C.: USGAO.

Conroy, P.J., D.M. Hughes, and I. Wilson. 1995. *Demonstration of an Innovative Water Main Rehabilitation Technique: In Situ Epoxy Lining.* Denver, Colo.: AwwaRF.

Cromwell, J., G. Nestel, R. Albani, L. Paralez, A. Deb, and F. Grablutz. 2001. *Financial and Economic Optimization of Water Main Replacement Programs.* Denver, Colo.: AwwaRF and AWWA.

Cromwell, J.E., III, H. Reynolds, N. Pearson Jr., and M. Grant. 2002. *Cost of Infrastructure Failure.* Denver, Colo.: AwwaRF.

Dandy, G.C., and M. Engelhardt. 2001. Optimal Scheduling of Water Pipe Replacement Using Genetic Algorithms. *J. Water Res. Pl.-ASCE,* 127(4):214–223.

Deb, A.K. 2003. Water Main Renewal, Planning and Prioritization. Presented at Distribution System Infrastructure Integrity Workshop, April 3–4, at Colorado State University, Fort Collins.

Deb, A.K., F.M. Grablutz, Y.J. Hasit, J.K. Snyder, G.V. Loganathan, and N. Agbenowski. 2002a. *Prioritizing Water Main Replacement and Rehabilitation.* Denver, Colo.: AwwaRF.

Deb, A.K., Y.J. Hasit, and F.M. Grablutz. 1995. *Distribution System Performance Evaluation.* Denver, Colo.: AwwaRF.

Deb, A.K., Y.J. Hasit, F.M. Grablutz, and R.K. Herz. 1998. *Quantifying Future Rehabilitation and Replacement Needs of Water Mains.* Denver, Colo.: AwwaRF.

Deb, A.K., Y.J. Hasit, and C. Norris. 1999. *Demonstration of Innovative Water Main Renewal Techniques.* Denver, Colo.: AwwaRF.

Deb, A.K., Y.J. Hasit, H.M. Schoser, and J.K. Snyder. 2002b. *Decision Support System for Distribution System Piping Renewal.* Denver, Colo.: AwwaRF.

Deb, A.K., Y.J. Hasit, J.A. Williams, and R. Jacob. 2001. *New Techniques for Precisely Locating Buried Infrastructure.* Denver, Colo.: AwwaRF.

Deb, A.K., K.A. Momberger, Y.J. Hasit, and F.M. Grablutz. 2000. *Guidance for Management of Distribution System Operation and Maintenance.* Denver, Colo.: AwwaRF.

Deb, A.K., J.K. Snyder, J.J. Chelius, J. Urie, and D.K. O'Day. 1990. *Assessment of Existing and Developing Water Main Rehabilitation Practices.* Denver, Colo.: AwwaRF.

Dingus, M., J. Haven, and R. Austin. 2002. *Noninvasive, Nondestructive Assessment of Underground Pipelines.* Denver, Colo.: AwwaRF.

Dolson, F.E. 1955. Factors Causing Main Failures. *Jour. AWWA,* 47(2):465–469.

Doyle, G., and M. Grabinsky. 2003. Applying GIS to a Water Main Corrosion Study. *Jour. AWWA,* 95(5):90–104.

Drangert, J., M.C. Nelson, and H. Nilsson. 2002. Why Did They Become Pipe-Bound Cities? Early Water and Sewerage Alternatives in Swedish Cities. *Public Works Manage. Policy,* 6(3):172–185.

Duranceau, S.J., M.J. Schiff, and G.E.C. Bell. 1996. *Effects of Electrical Grounding on Pipe Integrity and Shock Hazard.* Denver, Colo.: AwwaRF.

Edwards, M., S. Reiber, and W.D. Schecher. 1997. *Predicting Pb and Cu Corrosion By-Product Release Using CORRODE Software.* Denver, Colo.: AwwaRF.

Electronic Exhibit Hall on Safe Digging. N.d. [Online]. Available: <http://www.underspace.com/ex/safedign.htm>. [cited July 1, 2002].

Ellison, D. 2001. *Distribution Infrastructure Management: Answers to Common Questions.* Denver, Colo.: AwwaRF.

ENR (Engineering News-Record). 2001. Drinking Water: Replacement Costs Loom Large. *ENR,* 247(1):12.

Fanner, P. 2003. Non-Revenue Loss Reduction in the UK. Presented at Distribution System Infrastructure Integrity Workshop, April 3–4, at Colorado State University, Fort Collins.

Farley, M., and S. Trow. 2003. *Losses in Water Distribution Networks: A Practitioner's Guide to Assessment, Monitoring, and Control*. London: IWA Publishing.

Ferry, S. 2002. Discovering Why Pipes Fail. *Plumbing Engineer*, 30(8):31–36.

Frontinus, S.J. (AD 97). 1973. *The Two Books on the Water Supply for the City of Rome*. Translated by C. Herschel. Republished by New England Water Works Association, Boston, 1973.

Galleher, J. 2003. PCCP—Lessons Learned. Presented at Distribution System Infrastructure Integrity Workshop, April 3–4, at Colorado State University, Fort Collins.

Goodson, A.F. 2002. Utilization of Reliability Centered Maintenance. Presented at 2002 AWWA Infrastructure Conference, Chicago, Ill., March 10–13, 2002.

Gould, T.A. 2003. How MRI Works. HowStuffWorks, Inc. [Online]. Available: <http://electronics.howstuffworks.com/mri.htm>. [cited June 12, 2003]

Goulter, I., T.M. Walski, L.W. Mays, A.B. Sakarya, F. Bouchart, and Y.K. Tung. 2000. Reliability Analysis for Design. In *Water Distribution Systems Handbook*. Edited by L.W. Mays. New York: McGraw-Hill.

Grigg, N.S. 1996. *Water Resources Management: Principles, Regulations, and Cases*. New York: McGraw-Hill.

Grigg, N.S. 2002. *Water, Wastewater, and Stormwater Infrastructure Management*. Boca Raton: CRC Press/Lewis Publishers.

GTI (Gas Technology Institute). 2004. [Online]. Available: <http://www.gastechnology.org>. [cited January 14, 2004]

Guerrero, P.F. 2002. *Pipeline Safety: Status of Improving Oversight of the Pipeline Industry.* GAO-02-517T. Washington, D.C.: USGAO.

Gummersheimer, D.A., and J.W. Heavens. 1998. Evaluation of Two Novel Techniques for the Structural Renovation of Water Mains. Presented at AWWA Missouri Section Annual Conference, April 1998, Lake of the Ozarks, Mo.

Habibian, A. 2003. Pipe Rehabilitation Planning. Presented at Distribution System Infrastructure Integrity Workshop, April 3–4, at Colorado State University, Fort Collins.

Hafskjold, L.S. 2003. CARE-W: Computer-Aided REhabilitation of Water Networks. Presented at Distribution System Infrastructure Integrity Workshop, April 3–4, at Colorado State University, Fort Collins.

Hahn, M.A., R.N. Palmer, M.S. Merrill, and A.B. Lukas. 2002. Expert System for Prioritizing the Inspection of Sewers: Knowledge Base Formulation and Evaluation. *J. Water Res. Pl.-ASCE,* 128(2):121–129.

Harberg, R.J. 1997. *Planning and Managing Reliable Urban Water Systems*. Denver, Colo.: AWWA.

Heavens, J.W., and J.E. Gumbel. 2002. To Dig or Not to Dig: Design, Specification and Selection Issues in the Trenchless Renovation of Water Mains. In *Assessing the Future: Water Utility Infrastructure Management*. Edited by D.M. Hughes. Denver, Colo.: AWWA. 449–470.

Hellier, C.J. 2001. *Handbook of Nondestructive Evaluation*. New York: McGraw-Hill.

Herz, R.K. 1996. Aging Processes and Rehabilitation Needs of Drinking Water Distribution Networks. *Aqua J. Water Supp. Res. Technol.*, 45(5):221–231.

Herz, R.K. 1998. Exploring Rehabilitation Needs and Strategies for Water Distribution Networks. *Aqua J. Water Supp. Res. Technol.*, 47(6):275–283.

Holley, M. 1999. *Continuous Acoustic Nondestructive Evaluation of Prestressed Concrete Structures.* Pure Technologies, Inc. [Online]. Available: <http://www.soundprint.com/about_soundprint/technical/pipe5.pdf>. [cited June 12, 2003]

Hope, R. 2003. Establishing the Estimated Annual Replacement Rate for Water Mains. Presented at Workshop for Infrastructure: Above and Below Ground, AWWA Annual Conference and Exposition, Anaheim, Calif., June 15–19, 2003.

Hudson, W.R., R. Haas, and W. Uddin. 1997. *Infrastructure Management: Design, Construction, Maintenance, Rehabilitation, Renovation*. New York: McGraw-Hill.

Hughes, D.M. 2002. *Assessing the Future: Water Utility Infrastructure Management*. Denver, Colo.: AWWA.

Hunaidi, O., W. Chu, A. Wang, and W. Guan. 1999. Leak Detection Methods for Plastic Water Distribution Pipes. Denver, Colo.: AwwaRF.

Jackson, R.Z., C. Pitt, and R. Skabo. 1992. *Nondestructive Testing of Water Mains for Physical Integrity*. Denver, Colo.: AwwaRF.

Jacobs, P., and I. Goulter. 1991. Estimation of Maximum Cut-Set Size for Water Network Failure. *J. Water Res. Pl.-ASCE,* 117(5):588–605.

Jeyapalan, J.K. 2000. Growing Pipeline Market in America: 20 Billion Dollars for 2000. In *Pipeline Technology*, Vol. 1. Edited by R. Denys. Amsterdam: Elsevier Science B.V. 39–54.

Jordan, J.K. 2000. *Maintenance Management for Water Utilities*. Denver, Colo.: AWWA.

Karaa, F.A., D.H. Marks, and R.M. Clark. 1987. Budgeting of Water Distribution Improvement Projects. *J. Water Res. Pl.-ASCE,* 113(3):378–391.

Kennedy, J.L. 1993. *Oil and Gas Pipeline Fundamentals*. 2nd ed. Tulsa, Okla.: PennWell Publishing Co.

Kirksey, W.E. 2003. Pipeline Systems R&D for Critical Infrastructure Protection. Presented at ASCE Pipeline Conference, Baltimore, Md., August 17–20, 2003.

Kirmeyer, G.J., G.R. Boyd, N.K. Tarbet, and R.F. Serpente. 2000. *Lead Pipe Rehabilitation and Replacement Techniques*. Denver, Colo.: AwwaRF.

Kirmeyer, G.J., M. Friedman, K. Martel, and D. Howie. 2001. *Pathogen Intrusion Into the Distribution System*. Denver, Colo.: AwwaRF.

Kirmeyer, G.J., M. Friedman, K. Martel, and G. Thompson. 2002. *Guidance Manual for Monitoring Distribution System Water Quality*. Denver, Colo.: AwwaRF.

Kirmeyer, G.J., W. Richards, and C.D. Smith. 1994. *An Assessment of Water Distribution Systems and Associated Research Needs*. Denver, Colo.: AwwaRF.

KIWA. 2001. Research Summary: Clarity in Water Mains System Management [Online]. KIWA Web page. Available: <http://www.Kiwa.nl>. [cited September 20, 2003]

Kleiner, Y., and R. Balvant. 1999. Using Limited Data to Assess Future Needs. *Jour. AWWA*, 91(7):47–61.

Kwong, J.K.P. 2001. Innovative Tunneling Approach Used for Downtown Honolulu Sewer Restoration. *Public Works*, 132(7):32–35.

Loganathan, G.V., S. Park, and H.D. Sherali. 2002. Threshold Break Rate for Pipeline Replacement in Water Distribution Systems. *J. Water Res. Pl.-ASCE,* 128(4):271–279.

Male, J.W., and T.M. Walski. 1990. *Water Distribution Systems: A Troubleshooting Manual*. Chelsea, Mich.: Lewis Publishers.

Marston, A. 1913. The Theory of Loads on Pipes in Ditches and Tests of Cement and Clay Drain Tile and Sewer Pipe. Bulletin 31. Ames, Iowa: Iowa State University Engineering Experiment Station.

Matsushita, M. 2003. Performance of Kobe Distribution System in 1995 Kobe Earthquake. Presented at Distribution System Infrastructure Integrity Workshop, April 3–4, at Colorado State University, Fort Collins.

Mavrakis, G. 2003. Evaluating Condition of Water Facilities Using Predictive Life Cycles. Presented at Distribution System Infrastructure Integrity Workshop, April 3–4, at Colorado State University, Fort Collins.

Mays, L.W., ed. 1989. *Reliability Analysis of Water Distribution Systems*. New York: ASCE.

Mays, L.W. 2000. *Water Distribution Systems Handbook*. New York: McGraw-Hill.

Mays, L.W. 2002. Urban Water Infrastructure: A Historical Perspective. In *Urban Water Supply Handbook*. New York: McGraw-Hill.

McNeill, L.S., and M. Edwards. 2001. Iron Pipe Corrosion in Distribution Systems. *Jour. AWWA*, 93(7):88–100.

Mech-Chem Associates, Inc. 2002. Case Study: Armco, Inc. [Online]. Available: <http://www.mech-chem.com/bulk/armco.html>. [cited June 12, 2003]

Mergelas, B., and X. Kong. 2001. *Electromagnetic Inspection of Prestressed Concrete Pressure Pipe*. Denver Colo.: AwwaRF.

Moak, L.L., and A.M. Hillhouse. 1975. *Concepts and Practices in Local Government Finance*. Chicago: Municipal Finance Officer's Association.

Morris, R.E., Jr. 1967. Principal Causes and Remedies of Water Main Breaks. *Jour. AWWA*, 59(7):782–798.

Moruzzi, L. 1986. Reasons for Pipe Damage. Special Subject 16 (SS16-1–SS16-8). Presented at 16th International Water Supply Congress and Exhibition, November 2–7, 1986.

Moser, A.P. 2001. *Buried Pipe Design*. 2nd ed. New York: McGraw-Hill.

Moser, A.P., and K.G. Kellogg. 1994. *Evaluation of Polyvinyl Chloride (PVC) Pipe Performance*. Denver, Colo.: AwwaRF.

National Council on Public Works Improvement. 1988. *Fragile Foundations: A Report on America's Public Works*. Final Report to the President and Congress. Washington, D.C.: National Council on Public Works Improvement.

National Resource Council Canada, Institute for Research in Construction. 2002. Urban Infrastructure Rehabilitation Program. [Online]. Available: <http://irc.nrc-cnrc.gc.ca/uir/index.html>. [cited October 15, 2003]

National Utility Locating Contractors Association. 2004. [Online]. Available: <http://www.nulca.org/>. [cited July 3, 2002]

Neve, A.R., and M.R. Hedges. 1998. The Portsmouth Water Approach to Network Renewals and the Influence Upon Leakage. *Pipes and Pipelines Intl.*, 46(5):5–12.

Nikonow, P. 2001. Estimating Unavoidable Leakage in a Water Distribution System. Master's thesis. Colorado State University, Fort Collins.

O'Day, D.K., R. Weiss, S. Chiavari, and D. Blair. 1986. *Water Main Evaluation for Rehabilitation/Replacement*. Denver, Colo.: AwwaRF and USEPA.

Peterson, R.H. 1994. *Accounting for Fixed Assets*. New York: John Wiley & Sons.

Physical Acoustics Corporation. 2003. Acoustic Emission Technology. Mistras Holdings Group. Copyright 1995–2003 [Online]. Available: <http://www.pacndt.com/pacae.html>. [cited June 12, 2003]

Pierson, G., K. Martel, and A. Hill. 2001. *Practices to Prevent Microbiological Contamination of Water Mains*. Denver, Colo.: AwwaRF.

Pikus, I. 2003. Contaminant Monitoring. Presented at Methodology and Characteristics Workshop, June, in Philadelphia, ASCE/EWRI meeting.

Pramanik, A., L. Blankenship, and J.C. Moeller. 2003. Overview of WERF's Collection System Related Research Activities. Presented at ASCE Pipeline Conference, Baltimore, Md., August 17–20, 2003.

PRCI (Pipeline Research Council Intl.). 2002. *Technology for Energy Pipelines*. (Pipeline Operations, In-Service Testing and Repair Technology.) An eBook. Arlington, Va.: PRCI.

Price, R.E. 1990. The Investigation, Cause and Prevention of PCCP Failures. In *1990 AWWA Annual Conference Proceedings*. Denver, Colo.: AWWA. 663–683.

Public Works. 2001. Historic Community Uses Sliplining to Halt Sewer Infiltration. *Public Works*, 132(3):52–56.

Radovanovic, R. 2003. Critical Water Mains: The Need for Risk Analysis. Presented at Distribution System Infrastructure Integrity Workshop, April 3–4, at Colorado State University, Fort Collins.

Rajani, B. 2003. Condition Assessment of Water Mains: Approach for the Prioritization of Water Main Replacement. Presented at AWWA Annual Conference and Exposition, Anaheim, Calif., June 15–19, 2003.

Rajani, B., J. Makar, S. McDonald, C. Zhan, S. Kuraoka, C. Jen, and M. Viens. 2000. *Investigation of Grey Cast Iron Water Mains to Develop a Methodology for Estimating Service Life*. Denver, Colo.: AwwaRF.

Reed, G.D., and R.M. Tucker. 1988. *Evaluation of a Pilot Versus Field Distribution Corrosion Monitoring System*. Denver, Colo.: AwwaRF.

Reiber, S., S. Poulsom, S.A.L. Perry, M. Edwards, S. Patel, and D.M. Dodrill. 1997. *A General Framework for Corrosion Control Based on Utility Experience*. Denver, Colo.: AwwaRF.

Ross, S.A., R.W. Westerfield, and B.D. Jordan. 2000. *Fundamentals of Corporate Finance*. 5th ed. New York: McGraw-Hill.

Russell, B. 2003. CML Planning/Contracting at the Louisville Water Company. Presented at Distribution System Infrastructure Integrity Workshop, April 3–4, at Colorado State University, Fort Collins.

Seargeant, D. 2003. Water Main Rehabilitation: Prioritization. Presented at AWWA Annual Conference and Exposition, Anaheim, Calif., June 15–19, 2003.

Selig, E.T., and W.A. Nash. 1988. Buried Pipe Research Needs. In *Pipeline Infrastructure*. Edited by B. Bennett. New York: ASCE.

Selvakumar, A., R.M. Clark, and M. Sivaganesan. 2002. Costs for Water Supply Distribution System Rehabilitation. *J. Water Res. Pl.-ASCE,* 128(4):303–306.

Shorney, F.L. 2000. Forces Transforming the Water Business. *Jour. AWWA*, 92(1)88–93.

Siedler, M. 1982. Obtaining an Analytical Grasp of a Water Distribution System. *Jour. AWWA*, 74(12):628–630.

Singer, P. 2003. Future Issues of Water Supply Systems. Presented at AWWA Annual Conference and Exposition, Anaheim, Calif., June 16, 2003.

Smith, L.A., K.A. Fields, A.S.C. Chen, and A.N. Tafuri. 2000. *Options for Leak and Break Detection and Repair of Drinking Water Systems*. Columbus, Ohio: Battelle Press.

Spangler, M.G. 1941. The Structural Design of Flexible Culverts. Bulletin 153. Ames, Iowa: Iowa State University Engineering Experiment Station.

Stahre, P., J. Adamsson, and Ö. Eriksson. 2000. *Performance Indicators for the Water Industry: An Introduction*. VA-FORSK Report 2000.8. Stockholm: AB Svenska Byggtjäns.

Steirer, M.A., S.M. Katz, and C.A. Griffin Jr. 1998. Public Plays Vital Role in San Diego's Effort to Upgrade and Repair Its Water System. In *1998 AWWA Annual Conference Proceedings, Vol. A: Management and Regulations*. Denver, Colo.: AWWA. 81–95.

Sterling, R. 2003. Rehabilitation of Drains and Sewers: Book Review. *Trenchless Technol.*, 12(6)30–32.

Stratus Consulting. 1998. *Infrastructure Needs for The Public Water Supply Sector*. Report for AWWA. Boulder, Colo.: Stratus Consulting.

Sundahl, A.-C.M. 1997. Geographical Analysis of Water Main Pipe Breaks in the City of Malmo, Sweden. *Aqua J. Water Supp. Res. Technol.*, 46(1):40–47.

Thomas, G.L. 2001. Operations and Maintenance. In *Facility Design and Management Handbook*. Edited by E. Teicholz. New York: McGraw-Hill.

Thompson, C., and D. Jenkins. 1987. *Review of Water Industry Plastic Pipe Practices*. Denver, Colo.: AwwaRF.

Thompson, D.M., and S.A. Weddle. 1992. *Water Utility Experience With Plastic Service Lines*. Denver, Colo.: AwwaRF.

Thornton, J. 2002. *Water Loss Control Manual*. New York: McGraw-Hill.

Topf, H.E., Jr. 1990. The Inside Story of Joint Repair. In *1990 AWWA Annual Conference Proceedings*. Denver, Colo.: AWWA. 657–662.

Troutt, J. 2003. Water Quality and Distribution System Integrity. Presented at Distribution System Infrastructure Integrity Workshop, April 3–4, at Colorado State University, Fort Collins.

Tucker, M., and A.R. Corriveau. 2003. Leveraging Legacy Maintenance Data and GIS to Optimize Infrastructure Replacement and Rehabilitation. Laurel, Md.: WSSC.

USDOT (U.S. Department of Transportation). 2002. Research Program [Online]. Available: <http://primis.rspa.dot.gov/rd/index.htm>. [cited July 1, 2002]

USEPA (U.S. Environmental Protection Agency). 1997. *Research Plan for Microbial Pathogens and Disinfection By-Products in Drinking Water* (EPA 600-R-97-122) [Online]. Available: <http://www.nal.usda.gov/fsrio/ppd/epa03.pdf>. [cited September 15, 2003]

USEPA (U.S. Environmental Protection Agency). 2001. *1999 Drinking Water Infrastructure Needs Survey: Modeling the Cost of Infrastructure*. EPA 816-R-005. Washington, D.C.: USEPA.

Van Dyke, T. 2003. Cleaning and Lining, Sioux Falls. Presented at Distribution System Infrastructure Integrity Workshop, April 3–4, at Colorado State University, Fort Collins.

VEWIN (Vereniging van Waterbedrijven in Nederland). 2001. *Reflections on Performance, 2000*. Amsterdam: VEWIN.

Von Huben, H. 1999a. Selecting a Valve and Pipe Locator. *Opflow*, 25(9):12–14.

Von Huben, H. 1999b. *Water Distribution Operator Training Handbook*. 2nd ed. Denver, Colo.: AWWA.

Wallace, L.P. 1987. *Water and Revenue Losses: Unaccounted-for-Water*. Denver, Colo.: AwwaRF.

Walski, T.M., and A. Pelluccia. 1982. Economic Analysis of Water Main Breaks. *Jour. AWWA*, 74(3):140–147.

Warren, I.C., S.D. Mason, and P.J. Conroy. 1990. In Situ Cement Mortar Lining—Operational Guidelines and Code of Practice. Source Document for the *Water Mains Rehabilitation Manual*. Swindon, U.K.: Water Research Centre.

Westerhoff, G.P., D. Gale, P.D. Reiter, S.A. Haskins, and J.B. Gilbert. 1998. *The Changing Water Utility: Creative Approaches to Effectiveness and Efficiency*. Edited by J.B. Mannion. Denver, Colo.: AWWA.

WIN (Water Infrastructure Network). 2001. Clean and Safe Water for the 21st Century [Online]. Available: <www.win-water.org>. [cited July 17, 2001]

Wolfe, T.F. 1946. How to Prevent Breaks in Cast-Iron Pipe. *Jour. AWWA*, 38(6):765.

Woodcock, M. 2002a. Briefing for Project Investigator at WSSC, March 2002.

Woodcock, M. 2002b. Force Main Condition Assessment or Archeology of Modern Times. PowerPoint Presentation.

Woodcock, M. 2003. Research on Pipe Breaks and Renewal. Presented at Distribution System Infrastructure Integrity Workshop, April 3–4, at Colorado State University, Fort Collins.

WSSC (Washington Suburban Sanitary Commission). 2002. *Water Main Condition Assessment, 1915–1998, Analysis and Recommendations*; and *Water Distribution System: A Collection of Statistics, Facts, and Comments*. Internal reports, provided by M. Woodcock during the project. Laurel, Md.: WSSC.

Zhang, J. 2003. Past, Present, and Future: An Overview of AwwaRF Pipeline-Related Infrastructure Reliability Research. Presented at ASCE Pipeline Conference, Baltimore, Md., August 17–20, 2003.

Zimbelman, D. 2003. NCWCD Experiences With PCCP. Presented at Distribution System Infrastructure Integrity Workshop, April 3–4, at Colorado State University, Fort Collins.

ABBREVIATIONS

A–C	asbestos cement
AE	acoustic emission
ASCE	American Society of Civil Engineers
AWWA	American Water Works Association
AwwaRF	Awwa Research Foundation
AWWSC	American Water Works Service Co., Inc.
BMP	best management practice
CA	condition assessment
CARE-W	Computer-Aided REhabilitation of Water Networks
CDE	controlled destructive evaluation
CERF	Civil Engineering Research Foundation
CIP	cast-iron pipe
CIPP	cured-in-place pipe
CIRIA	Construction Industry Research and Information Association
CML	cement–mortar lining
CMMS	computerized maintenance management system
CPP	concrete pressure pipe
DB	database
DBP	disinfection by-product
D&C	design and construction
DIP	ductile-iron pipe
EBMUD	East Bay Municipal Utility District
EM	electromagnetic
ENR	*Engineering News-Record*
EWRI	Environmental and Water Resources Institute
FS	fully structural
GASB	Government Accounting Standards Board
GASB 34	Government Accounting Standards Board Statement No. 34
GIS	geographic information system
GPR	ground-penetrating radar
GPS	global positioning system
GTI	Gas Technology Institute
HDPE	high-density polyethylene

in.	inch
IR	infrared
IWA	International Water Association
KANEW	modeling software used to predict future replacement needs
KIWA	Dutch water research association
km	kilometer
LADWP	Los Angeles Department of Water and Power
LCR	Lead and Copper Rule
lf	linear foot
m	meter
MFL	magnetic flux leakage
mgd	million gallons per day
mi	mile
MMIS	maintenance management information system
MMS	maintenance management system
MRI	magnetic resonance imaging
NDE	nondestructive evaluation
NDT	nondestructive testing
NS	nonstructural
NSF	National Science Foundation
O&M	operations and maintenance
PAN	priority action number
PB	polybutylene
PCCP	prestressed concrete cylinder pipe
PE	polyethylene
PM	preventive maintenance
PPBS	planning-programming-budgeting system
PRCI	Pipeline Research Council Intl.
PVC	polyvinyl chloride
RCM	reliability-centered maintenance
R&D	research and development
RES	research
RFEC	remote field eddy current
RFEC-TC	remote field eddy current transformer coupling
RFP	request for proposal
S&A	sonic and acoustic
SDWA	Safe Drinking Water Act
SS	semistructural

T&D	transmission and distribution
3Rs	repair, rehabilitation, and replacement
UAW	unaccounted-for water
UIMDSS	Utility Information Management Decision Support System
U.K.	United Kingdom
U.S.	United States
USDOT	U.S. Department of Transportation
USEPA	U.S. Environmental Protection Agency
USGAO	U.S. General Accounting Office
VEWIN	the Netherlands Water Works Association and Accenture
WARP	Water Mains Renewal Planner
WEF	Water Environment Federation
WF	workforce
WIN	Water Infrastructure Network
WSSC	Washington Suburban Sanitary Commission
yr	year